SpringerBriefs on Pioneers in Science and Practice

Texts and Protocols

Volume 11

Series editor

Hans Günter Brauch, Mosbach, Germany

For further volumes:
http://www.springer.com/series/11446
http://www.afes-press-books.de/html/SpringerBrief_PSP.htm
http://www.afes-press-books.de/html/SpringerBriefs_PSP_TP.htm

Lourdes Arizpe

Migration, Women and Social Development

Key Issues

 Springer

Lourdes Arizpe
Regional Multidisciplinary Research Centre
National Autonomous University
   of Mexico
Cuernavaca
Morelos
Mexico

The cover photo is from the author's personal photo collection

Copyediting: PD Dr. Hans Günter Brauch, AFES-PRESS, Mosbach, Germany

Style Editing: Dr. Serena Eréndira Serrano Oswald, Cuernavaca, Mexico

Language Editing: Michael Headon, Colwyn Bay, Wales, UK

ISSN 2194-3125   ISSN 2194-3133   (electronic)
ISBN 978-3-319-06571-7   ISBN 978-3-319-06572-4   (eBook)
DOI 10.1007/978-3-319-06572-4
Springer Cham Heidelberg New York Dordrecht London

Library of Congress Control Number: 2014937688

Printed on acid-free paper

Springer is part of Springer Science+Business Media (www.springer.com)

*For women who are building more
sustainable relations in their personal lives,
in their societies and in the world*

Lourdes Arizpe addressing UNESCO's 33rd General Conference in October 2005. Photo was taken by Ph. Sayah Msadek and permission was granted by UNESCO

Books by the Same Author Published by Springer

- Lourdes Arizpe, Cristina Amescua: *Anthropological Perspectives on Intangible Cultural Heritage*. Springer Briefs in Environment, Security, Development and Peace, vol. 6 (Cham–Heidelberg–Dordrecht–London–New York: Springer-Verlag, 2013).

- Lourdes Arizpe: *Lourdes Arizpe: A Mexican Pioneer in Anthropology*. Springer Briefs on Pioneers in Science and Practice No. 10 (Cham–Heidelberg–New York–Dordrecht–London: Springer-Verlag, 2014).

- Lourdes Arizpe: *Culture, Diversity and Heritage: Major Studies*. Springer Briefs on Pioneers in Science and Practice No. 12. Subseries Texts and Protocols No. 6 (Cham–Heidelberg–New York–Dordrecht–London: Springer-Verlag, 2015).

- Lourdes Arizpe: *Beyond Culture: Conviviability and the Sustainable Transition*. SpringerBriefs in Environment, Security, Development and Peace, vol. 13 (Cham–Heidelberg–Dordrecht–London–New York: Springer-Verlag, 2015).

- See also the websites on this and other books by Lourdes Arizpe: http://www.afes-press-books.de/html/SpringerBriefs_PSP_Arizpe.htm and http://www.afes-press-books.de/html/SpringerBriefs_ESDP06.htm.

With students of the National School of Anthropology and History of Mexico (ENAH) during a field-work trip. This photo taken by Aida Analco and is reproduced with permission of the photographer

Preface

When I began studying the migration of indigenous women to Mexico City in 1973, the trickle-out of migrants in rural areas was already beginning to reflect the failure of the implementation of 'trickle-down' theories of development to bring about balanced economic and social developments in different sectors and regions. Since then, such outflows have become massive in many countries in the world, so much so that migrations, women, indigenous peoples, and the assimilation of migrants in host countries have become international priority issues today. I worked in these research fields from the 1970s to the 1980s, trying to bind them together in a framework of 'social development', and by the 1990s it seemed to me that a deeper sense of cultural dislocation was having either a constructive or a destructive effect on the historical fabric of many societies around the world. As my research drew closer to the issues of environmental global change, I then tried to piece together the dichotomies created by scientific models—ecosystems/social systems, natural resources/human resources (population), economic growth/social development—by addressing them within a framework of 'social sustainability'. This book reflects an anthropologist's intellectual pathway through these theoretical and thematic concerns and the attempt to bridge the anthropologist's empirical location with the anthropologist's macro-scale world perspective. It is also a testimony to how a researcher from a developing country joined so many others in the emergence of social science research in countries in the South.

Movement has always intrigued me. In a sense it has become the central concern of my anthropological research, both in terms of geographical migrations, and of shifts in cultural traditions and intellectual paradigms. Why do people move? Why do people change their inherited way of thinking and feeling? And then, how does the human capacity to create culture and to live in convivial societies have to be transformed to align development to the arrow of time?

As I grew up in a multicultural setting, I was struck by the cleavages and inconsistencies within cultures and religions and by militancies whose claims visibly ran counter to what seemed to be reality. The movement of peoples from different geographies and cultures fascinated me, perhaps because both my parents were migrants coming from different cultures.[1] Also, it happened that my Indian nanny

---

[1] My father, born in Monterrey, Mexico, had travelled as regional CEO for an American company to Peru and Argentina before settling in Mexico City. My Swiss mother had lived in Switzerland and Peru before marrying my father and spoke six languages.

used to whisper hair-raising legends, half in Náhuatl (the indigenous language of central Mexico), tugging me towards a world filled with the spirits of water springs and trees, and animals that talked.[2] In contrast to this imaginative richness, I was pained by the brutal discrimination that was inflicted on indigenous peoples. When I found that anthropology would give me the opportunity of entering such worlds, I read ravenously about Meso-American, African, and Oceanic cultures. The London School of Economics, where I received my postgraduate degree, gave me a global outlook and, most importantly, the understanding that all social processes take place in political and economic contexts.

This volume brings together many articles and chapters of books that I wrote in different fields of knowledge but which were woven together by the need to understand, early on, the 'first principles' of social life that Imre Lakatosh inspiringly taught us to look for, and later on, the 'patterns of culture,' as Ruth Benedict would have it, which would explain why people make the choices they make. The policy counterpart of such a research program would be the debates on development, especially sustainable human development, a discussion, and conceptual evolution I was very fortunate to have been invited to participate in.

Since many of these texts are not available at present, I am grateful to Springer for providing me with the opportunity of publishing them together. I have agreed to republish these texts not without hesitation. Can one justify bringing back into the light essays which were written at a particular time, with a particular language, in the context of specific scientific and social debates which have now shifted in their perceptual weight and pre-eminence? Reading them through, however, made me realize that there is, at the same time, a meta-analysis that may be usefully applied to the shifts in ideas and intentions in social science in recent decades. I have been, in fact, surprised to find that many of the research discoveries and analytic lines of inquiry I worked on are still valid and shed light on how different processes and disjunctions have improved human well-being or, in some specific instances, have worsened in terms of the human prospect. These essays, then, open a window into how and why, while everything *seems* to be changing, so much has *not* changed.

Cuernavaca, November 2013   Lourdes Arizpe

---

[2] My nanny, Juana Hernández, had fled to Mexico City after she shot (without killing him) her husband's murderer in her native town of Huejotzingo, Puebla. For many years, she was unable to go back.

Acknowledgments

Many colleagues encouraged my work when I was at a crossroads and helped me develop an international understanding of the challenges that bind us together. First of all I would like to mention Helen Safa, Susan Sontag, and Iris Murdoch, as well as Veena Mazumdar, Chie Nakane, Achola Pala, Devaki Jain, Fatima Mernissi, Leila Takla, Noyleen Heyser, Wazir Karim, and so many others. For their companionship and riveting discussions I would like to thank Orlandina de Oliveira, Elizabeth Jelin, Carmen Diana Deere, Vania Salles, Viviane Brachet-Márquez, Margarita Velázquez, Patricia Fernández Kelly, and Wendy Harcourt, among many others. The funding for the research projects described in this book was provided by the National Autonomous University of Mexico, the Colegio de México, the Ford Foundation, the United Nations Development Programme, the World Bank, the International Labour Organization, the Rockefeller Foundation, the Fulbright-Hays Foundation, and the John D. Guggenheim Foundation, among other institutions.

This book is filled with people's lives because they were willing to talk to me about them and so my first recognition goes to all of them. Yet this book would not have been possible without the relentless editorial dedication of Dr. Hans Günter Brauch, the editor of this book series who also did the copy-editing, to whom I am greatly indebted. In writing these articles I was fortunate to have been able to train many young researchers, among them Margarita Velázquez Gutiérrez, Cristina Amescua, and Edith Pérez Flores to name only a few of countless others, who made fieldwork an interesting and at times merry endeavor. I would especially like to thank Dr. Serena Eréndira Serrano Oswald (Mexico) for style editing, Mr. Mike Headon (UK) for language editing, Dr. Johanna Schwartz, the editor with Springer-Verlag and Ms. Janet Sterritt the producer with Springer-Verlag in Heidelberg (Germany), and the whole production team at Springer Publishers in Chennai, Tamil Nadu, India. Many thanks also to Norma Angélica Guevara for burrowing in many archives to find texts and to the Publications Department of the *Regional Multidisciplinary Research Centre* (CRIM) for their assistance in the preparation of this book.

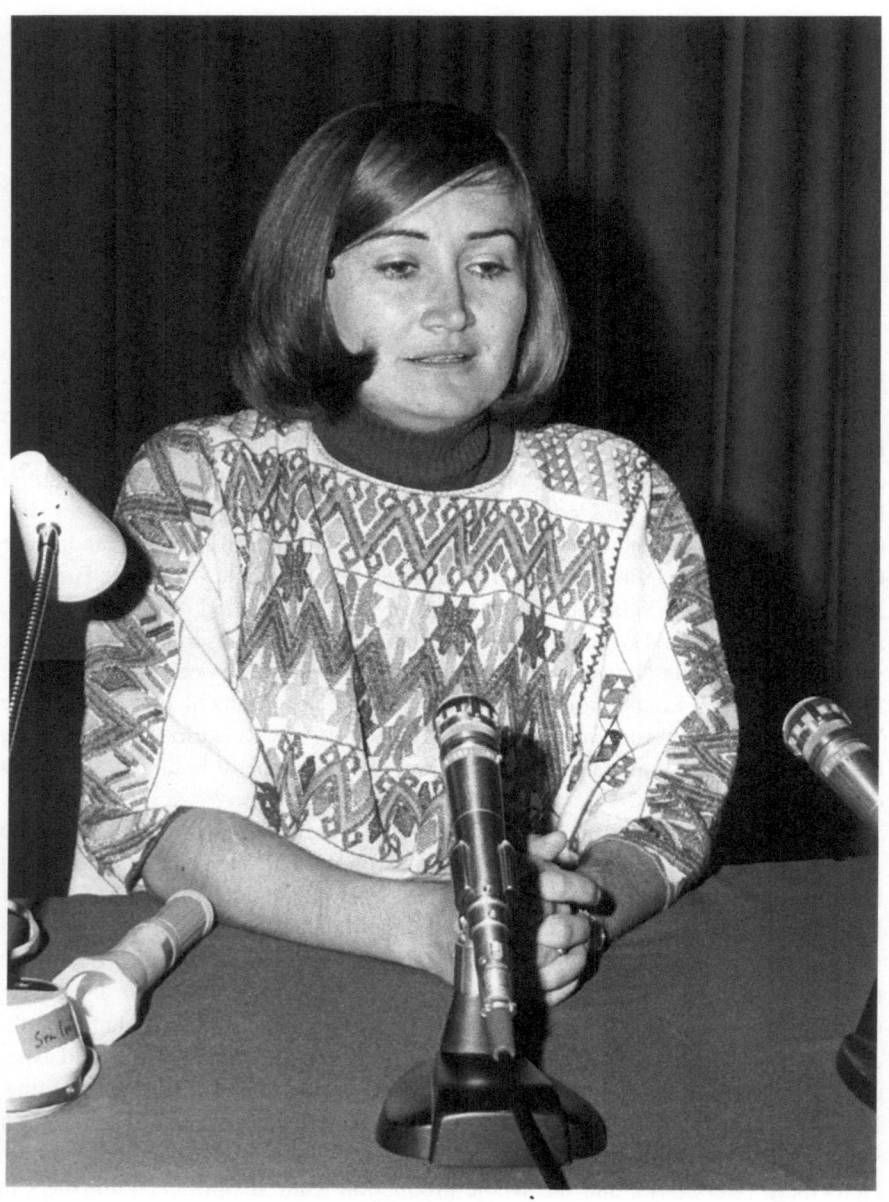

Lourdes Arizpe Schlosser at the UN Conference on Women in Nairobi in 1985. *Source* This photo was taken by Margarita Velázquez and reproduced with her permission

Contents

Reflecting about the indigenous dance of the Malinches in Tlacotepec in 2006. Photo was taken by Cristina Amescua and is reproduced with her permission

Chapter 1
Introduction

The phrase "the world is becoming a more dangerous place", now often heard in the voices of presidents, politicians and grass-roots activists, in the voices of Mexican migrants is tinged with a sense of incredulity: "all we want to do is work, why are we considered criminals?" For women migrants and children, especially, the dangers along migration routes have intensified as their families are torn asunder by insecurity, unemployment, and low-quality schooling. This compilation of my research looks back at these journeys and the bold ways in which people have tried to cope with risks at each bend in the road and it also looks forward to our journey towards sustainability. In addition, since the nineties, I had been writing about environmental sustainability and in 2010, I found I also had to write about how to restore social sustainability. This is the case with Mexico, certainly, but also, it seems to me, with many other developing and even industrialized countries. Part of the story of why this is happening today can be found in many of the texts included in this volume. As I kept repeating during this first decade, as a kind of mantra, when I was President of the International Social Science Council, "Policies must change: economies are growing, and societies are collapsing".

The first volume of the series on my research work, published by Springer (Arizpe 2014), focused on the findings of my anthropological and international research. In the present volume, the chapters provide an analysis that links local processes of development—rural change, migration, women's participation, movements such as feminism and cultural change—to the global issues of population, social development, and the human dimensions of global environmental change. The third volume (Arizpe 2015) will include recent texts on culture and development, intangible cultural heritage, and the anthropology of international policy-making.

1.1 Migration and Danger

Chapter 2 in Part I: Migration and Development explains why Mexican migrants now claim that their journeys to and from the United States have become more dangerous. While Mexican migrants and their communities have different perceptions

L. Arizpe, *Migration, Women and Social Development*,
SpringerBriefs on Pioneers in Science and Practice 11,
DOI: 10.1007/978-3-319-06572-4_1, © The Author(s) 2014

of racial and cultural groups in the US, they consider bilingualism (with English) and biculturality as a very probable future. Indeed, many studies in many countries have confirmed that bilingualism and biculturality are an asset when no discrimination or social tensions are present.

In Chapter 3, 'The Rural Exodus in Mexico and Mexican Migration to the United States', published in *The Border that Joins* (edited by Shue 1983), I describe how, in some regions of Mexico, migration has become a central strategy for rural development. Unlike the economic and sociological surveys of the time, which used questionnaires applied to individuals to establish the statistical frequency of migrations, my own fieldwork showed that the developmental cycle of domestic groups (which I had studied in my earlier research) led to a family strategy of migration that explained the selectivity of migrants by age, gender, order of siblings, and ethnic identity (Arizpe 2014, Chapter 7). Families sent out their daughters and sons in an age-ladder progression so they could send back cash remittances from the city. The causal model I proposed consisted of, instead of 'push-pull' mechanics, a first level of 'migrant motivations', a second meso-level of 'immediate causes' linked to the economic situation of the families, then a third macro-level of 'mediate causes' which reflected the impact of government economic policies and large-scale economic shifts.

At that time, when the fertility rate of rural women was very high, I predicted that rural outmigration in Mexico would become massive, although I could never have imagined, two and a half decades later, the colossal outmigration of more than seven million rural people resulting from the destruction of small-scale agriculture brought about by the North American Free Trade Agreement and the policies of the neo-conservative governments of 2000–2012 in Mexico.

1.2 Women Crossing Boundaries

In the last four decades women have crossed many boundaries—national, professional, psychological, symbolic—have pushed back boundaries—in work, knowledge, and mindfulness—and have also established boundaries—against brutality, against sexual harassment, against voicelessness—although new risks have arisen: new forms of economic violence, the regression of fundamentalist religions and the rise of might over right in many domains.

Part II: Women and Development includes texts from my early research on women's property and *ejido* rights in the Mexican agricultural system (Chapter 8), showing how women could be constantly displaced in such rights by kin or local authorities. Chapter 7 'The Comparative Advantages of Women's Disadvantages: Women Workers in the Strawberry Agribusiness in Mexico' demonstrates how, in spite of the 'disadvantages' faced by these young women from rural households in the strawberry factories—low wages, harsh health conditions, no medical or other services—they prefer this kind of work because it allows them to get out of their homes and to have greater autonomy. This text was included in an Oxford anthology on Women's Work and was translated into several Asian languages.

Since the eighties, after my ethnographic research had begun to focus on culture and development (Arizpe 1989), a John D. Guggenheim grant and funding from the Wenner-Gren Foundation has allowed me to extend this research interest to Peru, Brazil, India, Bangladesh and Senegal, and to other developing countries.

Soon after I had finished the research on these women workers, I started to participate in international research networks on women and development. I had been Secretary-General of the 'First Mexican-Central American Women Studies Conference' in 1977 which became, in fact, a Latin American and North American Conference. After that launch, and together with many initiatives in other countries, research into women's studies became a recognized scientific field while the social movement of feminism became a diversified political and intellectual trend that changed women's roles in society. In Chapter 6 I briefly describe the advances of feminist groups in Mexico, where I was very active from the seventies until the nineties, when I left Mexico to begin my international work on culture in Paris and Geneva.

In my research I returned to this field in 2011, when I wrote a paper for the Committee on Development Policy of the United Nations Economic and Social Council analysing the outcomes of so many women's movements across the world, especially the impacts on women's lives of development policies, cross-border migrations and illicit activities. This is included in Chapter 4, the 'Box on Migration, Gender and Global Crises' and Chapter 5, 'Women Crossing Boundaries', which I wrote on the basis of my participation as Chair of the Governing Board of the United Nations Institute for Research on Social Development (UNRISD) in Geneva.

1.3 A Society in Movement

In 1993, when organizing the International Congress of Anthropological and Ethnological Sciences, I coordinated a book called *Antropología Breve de México*, in which Mexican anthropologists of my generation summarized their main research findings. Chapter 9 is my own chapter in that text, and provides a broad canvas of social development in Mexico, beginning with the geo-historical mobility of ancient Meso-American cultures, as well as treating more recent indigenous movements, including the Chicano movement in the United States. The text spelled out the main challenges in terms of interdisciplinary anthropology and the urgency of facing concerns about globalization and the environment.

Almost twenty years later, in the midst of the worst stagnation and collapse of Mexican society under the neo-conservative governments, I wrote Chapter 10 'How to Restore Social Sustainability in Mexico'. At that time, the statistic calculated for deaths directly attributed to President Calderon's 'War against Drug Trafficking' was 35,000. By 2012, the count had risen to more than 100,000, without taking into account people who had disappeared. All other development indicators had decreased: according to data from the United Nations Economic

Commission for Latin America and the Caribbean (ECLAC; CEPAL in Spanish), Mexico was the country in the region with the highest growth in inequality, and the smallest gains against poverty. Drug production and trafficking had greatly expanded—after the head of the largest drug cartel, el 'Chapo' Guzman, had been allowed to escape from a maximum security prison and to figure currently in the list of the five hundred richest men in the world.

The concomitant impacts of such events in Mexico led to the highest rates of violence and to the lowest scores in the international Pisa educational surveys in recent history, as well as to increases in the number of youth suicides and massive outmigration to the US of people who would more aptly be called economic, social, and security refugees. In the text included here, I have suggested seven points to be implemented urgently, among them that Mexican unemployed youth should stop being murdered or imprisoned because they have become drug traffickers but rather that government policies should be restructured to turn this 'demographic bonus' (as in other countries) into a positive force for development.

1.4 The Human Dimensions of Sustainable Development

In the nineties, policies relating to migration, gender, and rural development quickly became enmeshed with debates about population growth and sustainability. In Chapter 11, 'The Social Dimensions of Population', a paper written for the Social Science Research Council of New York, Margarita Velázquez and I argued that understanding population growth meant going beyond statistical trends. Instead, such growth has to be analysed as a social process that is tightly influenced by local political and social trends and by cultural heritage. In fact, at that time, my participation in several international groups, especially one headed by Mahbub ul Haq, Richard Jolly, Francis Stewart, Sakiko Fukuda-Parr and others which led to the very important creation of the Human Development Report, thrust my work into the broader context of international social science.

While development was still not considered as having a 'human dimension', another intellectual scientific revolution was also taking off at the beginning of the nineties. In 1990, as President of the International Union of Anthropological and Ethnological Sciences, I began to actively participate in the first international initiatives to bring together the natural and social sciences in understanding and managing global environmental change. Chapter 12, 'Population and Natural Resource Use', which I wrote with Wolfgang Lutz, a demographer, and Robert Constanza, a biologist, is a chapter in the iconic book published by Cambridge University Press in which we all joined together from different disciplines, countries, genders, and perspectives to try to launch a new era of integrated science.

In Chapter 13, 'Human Dimensions of Global Change', I tried to consolidate a theoretical viewpoint in which the 'sociosphere', that is, the systems human beings create to live together in societies, could start to be taken into account as part of a heuristic approach in a way similar to the natural scientists' 'ecosystem'. However, such interdisciplinary work between the social and the natural sciences has fallen far behind in this twenty-first century.

Finally, Chapter 14 provides the ethnographic data on which I based the texts that I have just mentioned. In 1991, while absorbing the new knowledge related to global environmental change and sustainability, I realised that, in fact, there were very few micro-studies, whether in anthropology or in other social sciences, that could directly address the questions which were emerging in that new knowledge about environmental change and sustainability. The three-year research project I then plunged into in the Lacandón rainforest in south-east Mexico gave very interesting insights into local peoples' perceptions and conceptions of their environment and of their complex relationships to other groups involved in finding sustainable pathways towards the future.

As we are now all keenly aware, the world as constructed by our individual intellectual and political backgrounds is facing unprecedented challenges. As I look back at the research presented in these two volumes, I find a common thread: all the projects were aimed at finding out how people explain to themselves and to others what they do and why. In seeking the 'first principles' that my great teachers such as Imre Lakatosh taught me to look for in research, or the 'moral principles' that Iris Murdock told me held the world together, I always did find recognizable patterns in the groups and societies I was studying. Yet the moment I delved deeper into them, these patterns became constellations, each radiating towards different outcomes or sources. More is needed in order to understand them. Working with patterns will not lead us to sustainability. Constellations require different strategies, in which radiant procedures can build upon one another. The quest for knowledge, then, goes on. In a different sphere, yes. It has been my privilege to have explored so many worlds only to find, not the place where I started from, but, again, a new world.

Lourdes Arizpe
Cuernavaca, December 2013

References

Arizpe, Lourdes, 1983: "The Rural Exodus in Mexico and Mexican Migration to the United States", in: Brown, Peter G.; Shue, Henry (Eds.): *The Border that Joins: Mexican Migrants and US Responsibility*. Maryland Studies in Philosophy and Public Policy (city: Rowland & Middlefield): 162–183.

Arizpe, Lourdes, 1989: *Cultura y desarrollo: una etnografía de las creencias de una comunidad mexicana* (Mexico City: El Colegio de México–UNAM–Miguel Ángel Porrúa).

Arizpe, Lourdes, 2014: *Lourdes Arizpe Schlosser: a Pioneer in Mexican Anthropology* (Heidelberg: Springer Verlag).

Arizpe, Lourdes, 2015: *Culture, Heritage and Pluralism* (Heidelberg: Springer Verlag).

Interviewing migrants at the Mexico-US border crossing in 2012. This photo was taken by Armando Estrada who granted permission to reproduce it

Part I
Migration and Development

Interviewing a female worker in the strawberry agribusiness in Zamora in 1978. This photo is from the author's photo collection

Chapter 2
The World is Becoming a More Dangerous Place: Culture and Identity Among Mexican Migrants in the US

*Yes, I'd like to think of myself as an indigenous person,
they're very precious to me, but right now I feel normal–not
indigenous,
not Spanish, not a gringa–, I feel Mexican.
Rocío Montalvo, young migrant woman from Tlacotepec,
Mexico.*

2.1 Introduction

The second half of the twentieth century has been called 'the age of migration', given the unprecedented number of people who have moved to another country (Toro-Morn/Alicea 2004).[1] More than two hundred million persons—including undocumented workers—have crossed borders in their search for jobs or a new life. This is bringing about momentous changes in the hierarchies and creative mix of cultures, in governance structures and in human relationships.

Such flows are motivated mainly by disparities in revenues. Recent data in this report show that the gap in income between developed and developing countries tripled from 1960–1961 to 2000–2002: only 16 developing countries saw a growth of more than 3 % between 1985 and 2000, while another 55 % reported an annual growth of less than 2 %; of these, 23 reported negative growth (Halonen 2004).[2] Other factors are the disparities in population between countries and regions and, I would also emphasize, cultural influence of the mass media and audiovisuals.

The United Nations points out that the migration 'stock', which is the permanent nucleus of these transfers, rose from 0.8 % of the world population between 1960 and 1965, reached a surprising peak of 6.7 % in 1985–1990, and fell back to

[1] This text was originally published in *Development*, Vol. 50, Num. 4 (December 2007): 6–12. Permission granted by Ms. Laura Russell of Plagraves Macmillan on 2 July 2013.

[2] See also Global Commission on International Migration (GCIM) at: http://www.gcim.org (13 June 2013).

L. Arizpe, *Migration, Women and Social Development*,
SpringerBriefs on Pioneers in Science and Practice 11,
DOI: 10.1007/978-3-319-06572-4_2, © The Author(s) 2014

1.5 % in 2005–2007 (UN 2006).[3] The presence of women in these movements continues to grow, from 46.8 % in 1960 to 49.6 % in 2005 (UN 2006). These are no longer women traveling to join their families, they are independent workers; the same shift that has been identified in national migratory flows (for Mexico see Trigueros 2004; Goldring 2001).

Given globalization, geopolitical and demographic trends, recent estimates suggest that such flows will continue unabated. The illegal nature of much of international migration has also given rise to problems of crime and security that go beyond the border regions, which require new measures to combat trafficking in human beings, including the sexual commerce of women and children as well as organized crime and terrorism.

2.2 Cross-Border Migrations and Culture

The cultural aspects[4] of migration have rapidly become a salient issue in national and international agendas in relation to problems of migrants cultural integration or ascendancy, pluralism in governance and homegrown terrorism among migrants. Concerns about culture are currently being played out heatedly in three arenas.

First, in international relations where the old, discredited discourse of civilizations has been used, for example, example, by Samuel Huntington in the 'clash of civilizations', as a euphemism for religions, in what could be considered an attempt to simplify world politics so as to leave out the plurality of real actors presently emerging on the world scene.

Secondly, in the mass media and now in the Internet, in the concern over the dominance of the English language and of US audiovisuals in all realms of telecommunications and informatics.

How ironic it seems, then, that Huntington's 2005 book *Who are We? The Challenges to America's National identity*[5] calls for the defense of the 'American Creed' against Hispanic migrants in the US. He singles out Mexicans as the principal threat, arguing that they may end up dividing the United States into two

---

[3] Note that figures on international migration are based primarily on the number of foreign persons registered in national censuses and other official sources. Therefore, they do not include undocumented migrants, or trafficking, or those involved in the drug trade or other criminal activities.

[4] Culture being a polysemic term, the working definition of culture in this article is a self-referential group which usually belongs to broader categories, in which individuals decide whether to keep, discard or radically change the symbolic discourses and practices that are characteristic of the group. Cultures are not autonomous, autarchic units but are always linked by dimensional flows with other cultural clusters.

[5] A first article by Huntington (2004) presenting his argument was entitled 'The Hispanic Challenge' which Hispanics in the US quickly dubbed 'His Panic Challenge'.

nations. In fact, his data are wrong, since all studies show that by the third generation, descendants of migrations are fully integrated into the American way of life.[6]

Indeed, this is the third arena where culture has been true to its 'contentious' nature[7]: the integration or lack of integration of migrants in recipient societies. It is a policy area in which governments of countries receiving migrants are urgently looking for new policies, given the problems fostered by multiculturalist policies as in UK and Netherlands. While the problems of integration of migrants in developed countries are constantly referred to in international fora, much less attention is given in development policy discussions to the 'failed societies' that migrants are leaving behind them in their countries of origin.

In this context, Mexico occupies a particular place, because it is the only developing nation that shares a border with an industrialized country, a border that is also long (3,200 km long of which the Wall to be built would cover 700),[8] porous and politically sensitive. Demonstrations that brought out hundreds of thousands of undocumented workers of all nationalities and their allies in the United States in March and May 2006 drew attention to the need for political and cultural agreements in which the risks and benefits are shared more equitably between the sending and receiving societies.

2.3 Hispanic Immigrants in the US

According to the 2000 US Census, 22.3 million Mexicans are now living in that country, a number which doubled in 1980–1990 and again in 1990–2000 (Rumbaut 2006: 33) with an estimate that 4 million migrated between 2000 and 2006. In the census, that figure is equivalent to 63.3 % of the total Hispanic population of 35.2 million (Rumbaut 2006: 33).

How do different groups of immigrants identify themselves culturally, or, in the US Census terms 'racially'? According to Rumbaut (2006: 33), and I agree, Hispanics or Latinos are to being forced into racial categories that do not correspond to their historic experiences or cultural perceptions. Only 49.9 % of Hispanics considered themselves to be white, black or Asian; the rest identified themselves as 'multi-racial' or 'other'.

Mexicans classified themselves as follows: 46.8 % as 'white'; 0.7 % as 'black'; 0.2 % as 'Asian'; 1.2 % as 'American Indian'; 44.8 % as 'other', and the last 5.2 %

---

[6] In a recent article, data was presented from a survey showing the cultural impact of Mexican migrants as well as other Latino migrants, that has enlivened painting, literature, food and 'fiestas' in the US, and of course, i.e. popular music. See Arizpe (2004).

[7] The classic view of culture as a homogeneous, consensual, discrete unit has given way, since the rise of 'cultural studies' to perceiving it as a 'site of contestation'. This seems, indeed, a more apt definition of it for today's turbulent cultural interactions.

[8] At the Berlin Wall, 1,008 people died trying to cross. At the Mexican border, the count is at present in 2007 4,000. Migrants interviewed say the 'Wall of Shame' will not stop them because they have 'a passport from God'.

as 'two or more races' (Rumbaut 2006: 39). "Of those who chose 'other' as their race, 41 percent indicated Hispanic or Latino, and another 20 percent cited their nationality (e.g., Mexican)" (Portes/Rumbaut 2001: Table 7.7 cited in Tienda/Mitchell 2006: 44).

According to the excellent study carried out by the National Academies of Science on 'Hispanics' in the United States; figures and other social and cultural traits reveal the tremendous diversity among them (Tienda/Mitchell 2006). They explain that the term 'Hispanics', created by administrative decision as a new category for the 1980 census has, over various decades, become a new ethnic and racial category. Although it initially referred primarily to social and labor aspects, with the new movements that culminated in the massive demonstrations of May 2006, the category has taken on a strong political connotation.

The study also demonstrates that, although Hispanic immigrants in the United States are going through the same phases of immigration as immigrants of other nationalities—gradually losing their language and social isolation with each successive generation, attaining progressively higher levels of education and better-paying jobs, reaching professional and management positions in their companies and marrying with people of other ethnic groups, white or otherwise. They caution that unless more investment is devoted to the education and integration of Latinos, this category risks changing from an ethnic definition to a sort of permanent 'underclass', exacerbating the labour, educational, and health problems of this group (Tienda/Mitchell 2006: 57–58):

> The key question for the future is whether Latinos, the popular term for them, will evolve a symbolic identity for some or all people of Latin American descent as they join the American mainstream, or whether it will become an enduring marker of disadvantaged minority group status (Tienda/Mitchell 2006: 4).

On the other side of the border, our study shows how Mexican migrants themselves perceive their situation and the cultural interactions they are involved with in this continuous movement across the border.

2.4 How Do Mexican Migrants Identify Themselves?

In our fieldwork research[9] we interviewed migrants from four communities in the northeast state of Morelos and those of one community in the suburb of Norcross in Atlanta, Georgia. Migration in that region began only recently, with adult men migrating as temporary workers, although in the next generation, for today's

---

[9] Interviews and a survey of 201 individuals, based on a random sample with equal number of men and women, migrants and non-migrants, were carried out in the state of Morelos in Mexico and in one of the migrant communities in Norcross, a suburb of Atlanta, Georgia. They were conducted by Lourdes Arizpe, coordinator of the project, anthropology students Cristina Amescua and Edith Pérez, and Carlos Guadalupe Ocampo, a member of one of the communities. Work was funded by the National Autonomous University of Mexico and the Rockefeller Foundation.

young men and now young women between the ages of 15–30, migration is becoming a permanent option.[10]

Of the people interviewed, among the men, 64 % gave mainly economic motives for their migration, i.e.: "the farmland doesn't give enough", "there is no work", "they don't pay well here"; interestingly, another 15 % answered "to get to know the United States"—these being mostly young people—and 7 % gave as a motive that a relative had sent for them or had invited them to visit.[11] The answers by women were only slightly different: 52 % for economic reasons, 15 % to get to know the country and, as could be expected, 12.5 % at the invitation of relatives.

When asked with what country or region they most identified, respondents answered as follows: 76.1 % Mexico; 15.4 % the world; 7.0 % Latin America, and one respondent (0.5 %) mentioned the United States.

In terms of gender, it is significant that the number of women who identify first with Mexico is low (36.6 %) as compared to the number of men (61.4 %). In the in-depth interviews, women, in fact, voiced sharp criticism of Mexican institutions: of the government, because they blame it directly (as do their husbands) for forcing their husbands to emigrate, and for placing them in conditions of unemployment with no possibility of earning a decent income; of the courts and justice system, which has never offered them any protection against domestic or public violence—something homecoming migrants tell them is offered in the United States—indeed, with more than 400 violent deaths of women in Ciudad Juárez, among many other cases in other states of Mexico, unexplained by the Vicente Fox's government (2000–2006), together with the exponential growth in drug trafficking during that same period, these women are only describing reality; and, of social programs like *'Oportunidades'*—the World Bank sponsored social policy of the Mexican government—in which only a few women of the community are chosen to get a grant to keep their one eldest child in school. As one of the women left out of this scheme rightly said, "…they choose whoever they want to give the money to for whatever reason–what, so those of us who don't even get the dust, aren't we Mexican?"

One interesting aspect shown in the survey is that respondents who identify themselves as indigenous feel more closely identified with Latin America. Northeast Morelos still has a few people, mostly old women and men who speak 'Mexicano' as it is called locally, that is Náhuatl, the ancient language of the Toltecs and Mexica, representing no more than 5 % of the local population.[12] This being the case, it is highly significant that 56.7 % of those surveyed said that they did consider themselves indigenous! This points to how inclusive Mexican nationality is and that the term 'Mexicanidad' being used more and frequently by Mexican migrants and their descendants in the United States is, in fact, building a new social representation of them, broader than that of Mexican nationality, in the US.

---

[10] Fieldwork was conducted in northern region of the state of Morelos, in four communities in the municipalities of Temoac and Zacualpan, on the southern slope of the volcano Popocatépetl.

[11] The sample was random, but with a balance as to migratory condition (temporary, circulating or permanent), gender, age, and religion.

[12] Survey conducted in 2007 in four communities in the state of Morelos—Zacualpan, Tlacotepec, Amilcingo and Temoac—with a random sample of migrants and non-migrants cross-referenced for age, gender, locality and migrant status.

2.5 The World is Becoming More Dangerous

The most worrying trend indicated by the survey was the answers to the question in the survey: "what is the world's biggest danger today?" They describe an emerging world in which violence is predominating over all other risks, a fear expressed by 43.2 % of the respondents, with reference to crime, drug addiction,

Book cover of Lourdes Arizpe: *Peasants and Migration* (Mexico: Education Ministry 1985)

war, and terrorism. Other responses mentioned the environment (10.4 %), unemployment and poverty (7.0 %), illness (7.0 %), natural disasters (6.0 %) and other less frequent responses like migration, politics, human beings, and loose morals. Significantly, *twice as many women as men mentioned crime, and the response was the highest among young women between 19 and 35 years of age.* These migrants were especially emphatic in their remarks about crime and drugs, which reflect their experiences in crossing the border, and many of them living in high-crime, drug-plagued neighborhoods in the United States. They mention one specific aspect which, if corroborated, is highly troubling: "There are loads of people who have just given up, and don't come (back to Mexico)—they're into drugs". There are also increasing concerns about drugs expressed by a number of anxious mothers from Tlacotepec, in Morelos itself. A young boy from that town confirms the interpretation we made earlier on the deep insecurity that women feel living in Mexico. Even the two transsexuals who came up the sample of those interviewed in the US also identified crime as the biggest risk.[13]

Among the men interviewed—even the younger ones—there was also a predominant concern over crime and drugs. This was corroborated in interviews in which they said that, until recently, drugs were not being sold locally but are now circulating widely, even in the high schools of the region.

Explained to us, "the family gets lost when (the parents) go away, I guess because a lot of times the marriages don't last, they go away and they find someone else over there, and (the children here) are raised without an image, and they do whatever they want and do awful things because they don't have a strong father figure in their lives." This perception of an increase in drug consumption among children is correct: specialists at a seminar held by the Ministry of Health in May confirm that this type of consumption in Mexico rose at a rate of 20 % in the previous year.

Another prevalent response was about illness, also linked to the vulnerability they say they feel as immigrants. As a driver from Zacualpan said, "what's happening is there's a lot of infection from illnesses, deaths from the gases they inhale and they catch from each other … one of those was my brother, he got hit by a car over there … I'm more worried about the ones that went over, because they bring things back with them, over there they would give us talks about AIDS, and the doctor was a Mexican and said that he had a lot of cases of AIDS, and a lot of Mexicans go there, and how awful it felt to tell (his) countrymen that they had it … I can't forget that, because sometimes you just get yourself up and go, you can't get it from just one time, but think of their poor wives, because they come back here and infect their wives and your children, and now it's a pandemic". The problem is worsened by the lack of doctors or clinics that treat this sort of illness in their villages.

---

[13] Ethnographic fieldwork indicated that homosexuality may also be a factor in inducing migration, since the life of homosexuals in Mexican rural villages, both men and women, is wrought with discrimination and humiliation.

2.6 "I'm Thankful for Having Been Born in Mexico ..."

In the survey, cultural loyalties were laid out in interesting patterns. In response to the question of whether they preferred Mexican or US culture, 90.1 % of immigrants—most of them male—said Mexican. Only one chose American culture, and the rest did not answer or said they didn't know. We also asked why they preferred it, and we received an outburst of exuberant responses. Among them were the following: "I was in a (Mexican) dance group, and your blood swells up and you feel like you want to say 'Thank you for having been born in Mexico'" (Juan Martínez); or "Even though we're disrespectful and corrupt, looking at the warmth we create, I'm happy with that" (Rocío Montalvo).

There is however, bitter criticism against the Mexican border authorities, and some resentment against those who made it in the US. It is common that migrants complain that "...the same Mexican acts like a motherless jerk, they get some temporary job and get all full of themselves, and then when you get there they don't help you with anything, they discriminate against you" (Francisco Ornelas).

Another important finding was that most migrants like Americans, especially white Americans, many of them emphasizing how fair they are in treating them on the job. Migrants reiterate that "here (in Mexico) they give you a miserable wage that doesn't get you anything, sometimes it doesn't let you prosper; and (in the US) no, there if some employer sees you and you throw yourself into the job, they pick you up and then they even send for you".

Once again, though, women were more vocal in their criticism of Mexican culture: 97.1 % of the men said they preferred it, compared to 77.5 % of the women. The remainder (22.5 %) did not come out in favor of American culture; they merely did not answer or said they did not know. In the in-depth interviews, there were many instances in which the perceptions were expressed in criticism: against machismo, expressed in domestic violence and alcoholism, against the political disorder and against the irresponsibility of parents toward their children. Because of migration and the mass media, women now know that conditions are different for women in the United States.

It is interesting that 61.3 % of the migrants responded that they felt equally Mexican in Mexico as in the '*Norte*', but that 32.4 % said they felt more Mexican while in the US. Feeling 'more Mexican' fits in with a common attitude among migrants of different nationalities who find themselves in a any country having another lifestyle. However, Mexican identity is based on a multi-layered history of strife and nation-building as well as of notable cultural creativity going back two millennia. In fact, we found that all of these are present today in eastern Morelos: culture is permeated by the Mesoamerican Nahua heritage; some people still hold the agrarian ideology of the Mexican Revolution, even tinged with anarchist leanings, keeping up the flame of Emiliano Zapata's Zapatismo; new generations that have taken up the neo-Zapatismo of Subcomandante Marcos. At the same time, while several dioceses kept the fundamentalist Catholicism predating the Vatican Council II, in our survey, 21.2 % of respondents said they profess no religion.

This plural vision of the world the migrants' sites of origin in our study encounters a different plurality in the US: more contacts with other Latin American and Caribbean people, as well as the astonishing variety of Americans. Eleazar Martínez puts it this way: "it seems as if Mexicans are hot-blooded, party-goers, if you're just hanging out with someone you feel (good), but over there (in the United States) they don't have much culture".

2.7 Biculturality Without Problems

Ultimately, what matters is the future. What do migrants and non-migrants think of what will happen in the future? Let's take the example of the survey's question about whether Spanish will be spoken alongside English in Mexico in the future. A surprising 70 % said yes. A very sharp contrast is evident between young people having less than 35 years of age, among whom the answer was as high as 80 % while among those older than 36 years this figure sinks to around 50 %. There is also a marked difference between Catholics (58.55 %) and Protestants (88.9 %).

Although they may view this change as inevitable, attitudes toward this varied significantly: 48.9 % of respondents think it would be a good thing, 17.8 % a bad thing, and 12.2 % are indifferent to the prospect. The age difference is not so pronounced as in the previous question, but it does reflect a particular bias among teenagers (13–18 years old), 64.3 % of whom think it would be a good thing. These are precisely the age groups more eager to migrate to the United States—concrete evidence of this can be seen in the fact that 80 % of the male graduates of one high school class in Tlacotepec have gone to 'el Norte'.

Religion is another interesting factor in this response—61.1 % of Protestants approve of bilingualism, compared to 47.2 % of Catholics (15 % of which did not feel it was either good or bad) and 42.1 % of those who professed no religion.

2.8 Conclusion

Our fieldwork and survey data gave a vision of the plurality and contestation of cultural interactions through migration to the United States in the region that was studied. It also pointed at what may be considered to be a general characteristic of Mexican culture: its cultural generosity, which receives and accepts the co-existence of different cultures, ideologies and religions. It also nurtures cultural creativity and crystallizes in the richness of a constantly recreated humus of intangible cultural heritage. For many of the migrants who are crossing the border, however, this situation becomes very different in the United States.

Over there, they say that they feel that there are no sets of practices, rituals or festivities that would allow them to reconstruct the plurality and density of social, family, ritual kinship and community relations that they enjoy in Mexico. It is cause for great concern that, if undocumented or illegal Mexican migrants tend to get sucked into marginal or even dark criminal undergrounds—even more so at present with the threats of deportation as shadows of a recession loom—and as they become represented in some US media in a dumbed-down version of Mexicans, they face a cultural loss of plurality and possibilities to display their heritage of cultural creativity. So they may turn even more militantly to their identity and heritage.

This turns Samuel Huntington's hypothesis upside-down: if he sees a threat of Mexican dividing the US, it will become more exacerbated the more Mexicans are pushed into a lowly status as a permanent underclass, together with other Latinos.

Cross-border migrations the world over are overwhelming all previous policies and cultural fences meant to melt them, mix them or keep them apart from their host societies. We urgently need new formulas in development thinking to recast cultural representations, multilayered loyalties, in a world in which nations are far from melting down.

References

Arizpe, Lourdes, 1983: "The Rural Exodus in Mexico and Mexican Migration to the United States", in: Brown, Peter G.; Shue, Henry (Eds.): *The Border that Joins: Mexican Migrants and US Responsibility* (Maryland: Institute of Philosophy and Public Policy): 162–183.

Arizpe, Lourdes, 2004: "Migración y cultura: las redes simbólicas del futuro", in: Arizpe, Lourdes (Coord.): *Los Retos Culturales de México* (Mexico City: Mexican Senate—UNAM—Miguel Ángel Porrúa): 19–42.

Arizpe, Lourdes, 2006: *Culturas en movimiento: interactividad cultural y procesos globales* (Mexico City: Mexican Senate—Miguel Ángel Porrúa).

Delgado Wise, Raul; Favela, Margarita (Eds.), 2004: *Nuevas tendencias y desafíos de la migración internacional México-Estados Unidos* (Mexico City: Mexican Chamber of Deputies—Miguel Ángel Porrúa—Universidad Autónoma de Zacatecas).

Durand, Jorge; Massey, Douglas, 2003: *Clandestinos: migración México-Estados Unidos en los albores del siglo XXI* (Mexico City: Miguel Ángel Porrúa).

Fernandez-Kelly, Patricia, 1995: "Social and Cultural Capital in the Urban Ghetto: Implications for the Economic Sociology of Immigration", in: Portes, Alejandro (Ed.): *The Economic Sociology of Immigration* (New York: Russell Sage Foundation): 213–247.

Goldring, Luin, 2001: "The Gender and geography of citizenship, Mexico, US: Transnational Spaces", in: *Identities*, 7, 4: 501–537.

Halonen, Tarja, 2004: "Toward a Fair Globalization: A Finnish Perspective", paper for the U Thant Lecture Series, 8th Lecture, Commissioner of the World Commission on the Social Dimension of Globalization, United Nations University, Tokyo, 22nd October.

Huntington, Samuel, 2004: "The Hispanic Challenge", *Foreign Policy* (March 1), at: http://www.foreignpolicy.com/articles/2004/03/01/the_hispanic_challenge (13 June 2013).

Huntington, Samuel, 2005: *Who are we? The Challenges to America's National Identity* (New York: Simon & Schuster).

Imaz, Cecilia, 2004: "Las Organizaciones sociales de migrantes mexicanos en Estados Unidos", in: Delgado Wise, Raul; Favela, Margarita (Eds.): *Nuevas tendencias y desafíos de la migración internacional México-Estados Unidos* (Mexico City: Mexican Chamber of Deputies—Miguel Ángel Porrúa—Universidad Autónoma de Zacatecas): 47–67.

Kearney, Michael, 1995: "The Local and the Global: the Anthropology of Globalization and Transnationalism", in: *Annual Review of Anthropology*, 24: 547–565.

Portes, Alejandro; Rumbaut, Rubén, 2001: *Legacies: The Story of the Immigrant Second Generation* (Berkley: University of California Press).

Rumbaut, Rubén G., 2006: "The Making of a People", in: Tienda, Marta; Mitchell, Faith (Eds.): *Hispanics and the Future America* (Washington, DC: National Academies Press): 16–66.

Tienda, Marta; Mitchell, Faith (Eds.), 2006: *Hispanics and the Future America* (Washington, DC: National Academies Press).

Toro-Morn, Maura; Alicea, Marixa (Eds.), 2004: *Migration and Immigration: a Global View* (Westport, Ct: Greenwood).

Trigueros Legarreta, Paz, 2004: "La migración femenina mexicana hacia los EUA y su participación en el mercado laboral de ese país", in: Delgado Wise, Raul; Favela, Margarita (Eds.): Nuevas tendencias y desafíos de la migración internacional México—Estados Unidos (Mexico City: Mexican Chamber of Deputies–Miguel Ángel Porrúa–Universidad Autónoma de Zacatecas): 97–127.

United Nations, 2006: *Trends in the Total Migrant Stock: the 2005 Revision* (New York: UN Population Division).

World Bank, 2005: Income generation and social protection for the poor (Washington, DC: World Bank).

Why are walls rising in a globalized world? Looking at the Mexico–US border Fence in Tijuana, 2012. This photo taken by Edith Pérez Flores and is reproduced with her permission

Chapter 3
The Rural Exodus in Mexico and Mexican Migration to the United States

3.1 Introduction

Migration out of rural areas, a pervasive movement in most developing countries since the fifties, began in Mexico after the turn of the century and greatly intensified in the last three decades.[1] During this same period, as Bustamante's (1977) and García's (1983) historical research has shown, labour conditions in the United States attracted Mexican migrants, mostly from rural areas, in sharply fluctuating patterns of active recruitment, laissez-faire, or repatriation. Because these two movements have varied simultaneously and because they are interrelated, it has been assumed that the rural exodus in Mexico generally explains the flow of migrants across the border to the United States. This chapter will argue that they must be analysed instead as two distinct movements. Data will be presented to show that most of the migrants created by the prevailing conditions in Mexican rural villages settle within Mexico, and that only specific types of migrants are attracted over the border.

This distinction between the general rural outflow in Mexico and Mexican migration to the United States is crucial for an understanding of who the US-bound migrants are in Mexican rural society. This is particularly necessary because, in spite of greatly improved data and research on communities sending migrants to the United States, the patterns, reasons, and rhythms of the movement

---

[1] This text was originally published as "Cultural Change and Ethnicity in Rural Mexico", in: David Preston (Ed.): *Environment, Society and Rural Change in Latin America* (London: John Wiley and Sons Ltd., 1980): 123–134 and as: "The Rural Exodus in Mexico and Mexican Migration to the United States", in: *International Migration Review*, XV, 4 (Winter 1981): 626–650. Permission was granted on 4 July 2013 by Ms. Verity Butler, Permissions Co-ordinator, Wiley, Chichester, UK.

L. Arizpe, *Migration, Women and Social Development*,
SpringerBriefs on Pioneers in Science and Practice 11,
DOI: 10.1007/978-3-319-06572-4_3, © The Author(s) 2014

still tend to become buried under broad generalizations that seek simple villains, such as population growth, unemployment, or poverty, for what is a highly complex social and economic process.

A finer knowledge of the texture of these two flows can be important in formulating policy for all sides concerned. Most policy proposals, with a few exceptions, seem to envisage uniform solutions to what is in fact, in the case of Mexican migration to the United States, a heterogeneous flow. Different types of migrants, displaying diverse strategies and targets, which for them involve difficult personal ordeals, require, if not special policies, at least an acknowledgement of their differences in any assessment of the impact of their migration on sending and receiving communities.

I begin by discussing some of the basic issues in the analysis of rural outmigration in developing countries. Then I examine the diversity of conditions that create potential outmigration in Mexican rural villages, as a context in which to understand which migrants are being attracted to the United States.

3.2 Rural Outmigration and Development: A Few Historical Hints

It is important to begin by stating that the experience of countries with early industrialization should incline us not to be surprised by rural outflows that lead to internal and international migrations. All industrial nations recruited the bulk of their industrial workforce from surplus agricultural labour that flowed to industrial regions where capital investments allowed greater productivity.

The rate at which labour flows out of agriculture, stimulated by multiple factors, usually outstrips the rate at which it is absorbed into the new industries. During the period when Europe was industrializing, for example, it was unable to absorb all the surplus labour being released from agriculture: between 1846 and 1932, it sent 51 million emigrants overseas (Thomas 1996), Each of the four great outflows of migrants from Europe between 1844 and 1913 was for the most part a rural exodus. As example, part of the outflow in 1849–1854 came from the break-up of the rural economy in the southwest of Germany; that of 1881–1888 was brought about by the agricultural crisis of that resulted from cheaper wheat imports from the United States (Thomas 1996: 10).

More than half these migrants were welcomed into the United States, which took in more than 32 million European immigrants between 1821 and 1932. Most came from rural hinterlands, as the US Immigration Commission stated for the 1903–1913 inflow of 10 million migrants. "Before coming to the US, the greater proportion was engaged in farming or unskilled labour and had no experience or training in manufacturing industry or mining" (Thomas 1996: 10).

The export of redundant labour in the European case had been made possible by the vast spaces opened overseas for European domination and colonization

Mexican migrants on their way to crossing the Mexico–US border 2012. This photo was taken by Edith Pérez Flores and is reproduced with her permission

during the previous two centuries. This option, of course, is not available for developing countries today.

Several features of rural outmigration are worth noting here. First, the outflow of rural migrants alluring Western Europe's industrializing period occurred in spite of slow population growth in its rural areas. This suggests that the high population increase in developing countries today cannot always be singled out as the unique cause of rural outmigration. Instead, this difference points to how much more critical a situation developing countries are facing today, where policies similar to those applied for Western European industrial growth are bringing about similar patterns of rural–urban migration, but with greatly increased population burdens and fewer employment or overseas migration options.

Second, blaming poverty as the main cause of rural outmigration explains very little. Poverty existed long before massive rural–urban migration began. Moreover, it is as much a symptom of underlying economic processes as rural outmigration is.

Third, ethnic differences, except in cases of political strife, invasion, or war, are usually reflected only indirectly in migration. Granted, the ethnic factor is crucial in channelling the distribution of land, wealth, and political influence, thus creating a pattern of class stratification that is reflected in the selectivity of migrants. Its influence is more direct, then, in the patterns of migration, since economic and social bonding strengthened by a common ethnic identity shape a distinctive behaviour in migrants of that particular group. This is especially true in the case of Mexico (Butterworth 1971; Arizpe 1978).

3.3 Understanding Migration

The close association between the growth of industrial capitalism, especially in countries with a highly centralized pattern of industrialization, and large-scale rural-urban migration suggests that this type of labour migration can be considered the geographical expression of an economic process. It would be inappropriate, however, to turn the truism that the rural exodus is generally the result of the development of capitalism into an abstract principle to explain migration. Such a view fails to make the theoretical distinction between the creation of a relative surplus population and the process of migration. Beyond doubt, the release of labour through higher levels of capital investment in agriculture is the necessary condition for massive outmigration. But it is another thing to say that the workers thus released automatically become migrants. Many alternative moves are possible. Workers may turn to other local salaried or income-generating activities; cultural and ethnic prescriptions may deter or change the selectivity of migrants; the labour redundancy may be transferred from one sex to the other—for example, young women of the household may be sent away to work to offset the deficit in the predominantly male agricultural activity, or young men may migrate if rural factories seeking low labour costs prefer to employ young women.

Thus, migration must be examined at two levels of analysis. At a general level, it must be understood within the context of industrialization, urbanization patterns, the transformation of peasant economy into a market economy, and state policies influencing economic and social change. These channelling forces shape migration, but this level of analysis can yield only broad, hypothetical explanations of migration. It is still necessary to explain why some rural inhabitants stay while others migrate, when they all face similar pressures to leave. This means that, since rural migrants are not a random sample of rural inhabitants, their *selectivity* must be explained.

"Feel safe yet?" Graffiti at the Mexico–US border Wall, 2012. This photo was taken by Lourdes Arizpe and is from the author's photo collection

Migrant selectivity can be explained only at another level of analysis, *by shifting the focus of the analysis from the migrants themselves, to what is happening in the social structure of rural towns and villages*. This implies bypassing their personal idiosyncrasies (whether they are psychologically the most adventurous, or the most passive, etc.) and the way they explain their reasons for migrating, and instead examining their position within the household (i.e., if they are the eldest or youngest offspring, male or female) and within the community (i.e., whether landowners wage labourers, craftsmen, traders).

The essential point here is that the characteristics of migrants become significant only within a given social structure. *It follows, then, that who migrates will almost certainly vary with each generation and, in even more cases, according to the pace of social change.* Rural outmigration thus can be fully accounted for only by looking closely at historical processes and specifically at the variations occurring in regions and communities undergoing rapid social change.

To summarize, there are two basic questions to be asked about rural outmigration in any developing country. The question of why migration occurs will be related to policies of agricultural and industrial development. But the particular features of migration can be explained only by asking how rural communities and individuals are coping with the pressures to migrate inherent in centralized industrial capitalism.

With these analytical instruments at hand, I will try to explain the major forces behind the rural exodus in Mexico. It would be beyond the scope of this essay to analyse fully Mexico's economic development of the last few decades. Therefore, the macroeconomic changes most relevant to explaining migration will be mentioned only briefly and analysis will focus on the changing social structure of rural villages.

3.4   Development Policies and the Rural Exodus in Mexico (1940–1980)

The drift from the land began long ago in the Mexican countryside—and even in the last century part of it involved migration to the United States—but it took on a definite pattern only in the early forties (García/Griego 1983). During the previous decade, President Cárdenas had carried out an agrarian reform that favoured rural peasants by redistributing land extensively. This set off an era of rural prosperity, which many consider the backbone of the successful 5.7 % annual growth of Mexican agriculture from 1940 to 1965. By the end of the sixties, however, agricultural growth had dropped below the level of the rural population increase, staple grains had to be imported, and a flood of migrants was coming to the cities and crossing the border. What had happened? What turned the production of food into the production of migrants?

In the beginning, the success of the agrarian reform and the favourable balance for agriculture in the national economy—the price of food increased more than the general price index between 1929 and 1945 (Solís 1970: 23–70)—were reflected in improved nutrition and greater access to medical services that lowered mortality rates and created an unprecedented population growth of more than 3 % per year.

Beginning with Aleman's regime in 1948, a new strategy of development was adopted, which directed government efforts and investments toward import substitution-substitution and large scale irrigation agriculture.[2] Although federal expenditure

---

[2] The import-substitution policy advocated that developing countries begin their industrialisation by producing commodities which they previously imported from abroad.

in agriculture grew progressively after that time, it was unevenly distributed within the agricultural sector (Bancomext 1976: 146–148). By 1960 the proportion of federal investment in irrigated agriculture was double that directed toward rain-fed agriculture, and by 1966 it was seven times as great (Bancomext 1976; Schumacher 1981).

Government policies encouraging rapid industrialization were at the same time fostering the growth of urban centres, which increasingly attracted people from rural regions. Employment was created at a very rapid rate in the major cities, particularly in Mexico City. There, for example, 503,000 jobs opened up in the forties, 686,000 in the fifties, and 679,000 in the sixties (Contreras 1972). Migrants could find formal employment easily since jobs had low entrance requirements, and they could be trained on the job. For those seeking temporary income, the construction boom of expanding housing and urban infrastructure offered ample opportunities of taking up and leaving jobs as needed. Not only Mexico City, but other cities like Guadalajara and Monterrey, although on a much more moderate scale, were also building up an industrial workforce from the influx of rural migrants.

The forties and early fifties, then, were the golden age for rural migrants in Mexico, the era that created the myth of plentiful work and money that continues to exercise a strong influence in fostering migration from rural villages in the hinterland. Because of conditions in both irrigated and small landholding agriculture, government policies were successful in providing cheap food for the cities, in shifting labour from agriculture to manufacturing industry, and in creating urban employment.

Nevertheless, in spite of the spectacular growth, agricultural development was very uneven, both in its exchange with the industrial sector and internally. Agriculture provided the exports and cheap food needed to set Mexico on the path toward industrialization, but in the process it lost too many resources. Through fiscal mechanisms, more money was taken from agriculture through taxes than was put back from 1940 to 1963 (Gómez 1978: 718).

The unevenness becomes most evident in the structure of prices. From 1940 to 1966, the net balance of the fluctuation of agricultural prices in relation to the general price index showed that 2,905 million pesos (at 1960 prices) were transferred to the rest of the economy (Gómez 1978: 718). In recent years, massive government investment in the agricultural sector has offset the long-term loss of resources so that, overall, the balance is now equal.

But it was the prolonged, fluctuating imbalance in the exchange between agriculture and industry, which transferred a total of 2.3 % of agricultural production between 1942–1962 that first began to erode the economic base of small farms (Reyes 1974). It particularly undermined small landowners because of the uneven development inside the agricultural sector.

The increased irrigation of arable lands benefited only a limited group of capitalist entrepreneurs and displaced smaller producers, and the Green Revolution further polarised this development, as it did in other developing countries (George/ Susan 1977). As Hewitt de Alcántara has explained for Mexico, the hybrid seeds could produce higher yields only when used with other high-technology inputs of insecticides, chemical fertilisers, and irrigation. Only a minority of farmers were able to benefit from the new strains and irrigated land. As a result, technology and capital began to concentrate in larger estates (Hewitt 1979).

By 1960, this uneven development of Mexican agriculture was evident in the following figures: 50.3 % of land plots had less than five hectares of land, which amounted to 13.6 % of the total arable land. These *minifundios* produced 4.2 % of agricultural production and owned 1.3 % of agricultural machinery. By contrast, the top 0.5 % of land plots took in 28.3 % of the arable land (including 37.6 % of irrigated lands), produced 32.2 % of agricultural production, and owned 43.7 % of agricultural machinery (Reyes 1974: 220).

But the hardest blow to the *minifundios*, which subsisted on rain-fed maize cultivation, came in 1957 when the price of maize was artificially regulated. Since 1925, the price of maize had risen steadily in cycles, but in 1957, on the basis of a policy of 'stabilising development' (*desarrollo estabilizador*), the rising price cycle was stopped with massive imports, a procedure that was repeated in 1963 when the price of maize again began to rise (Gómez 1987: 727). This regulating mechanism worked until 1969 because the increased yields of the Green Revolution allowed large farms to offset the loss in price. The government-guaranteed price of maize was maintained from 1957 until 1973, during which time the Mexican economist Gómez Oliver (1978) estimates that maize lost 33 % of its value.

This loss affected small rain-fed farms much more severely than large irrigated farms, and thus the former have felt the unevenness of development within agriculture and the transfer of resources to industry much more directly. Warman states that the balance between monetary costs of production, prices, and mean salaries in rural areas was broken in 1966 (Warman 1978: 681–687). *Minifundio* farmers then lowered their 'investment' in agricultural production and increased their dependence on finding seasonal wage labour. By 1974, two million hectares of rain-fed land, which had been cultivated in 1965, had been abandoned (Warman 1978: 681–687), hence the crucial shortage in staple grain production in Mexican agriculture in the seventies. When one also recalls the decline of wage employment in rural areas in this period, it becomes clear how dependent the *minifundio* farmers became on migratory wage labour. The farmers relied increasingly on seasonal and temporary labour in the cities and in the expanding agricultural regions in the Northwest.

Male and female wage labourers represented 36.7 % of the agricultural workforce in Mexico in 1950, 48.0 % in 1960, and 54.0 % in 1970 (Paré 1977). Of landless agricultural workers, 58 % live in the central-northern states of Puebla, Querétaro, San Luis Potosí, Tlaxcala, Morelos, Michoacán, México, Jalisco, Hidalgo, Zacatecas, Guerrero, and Oaxaca. These are the main regions of rain-fed agriculture. Many of these labourers work in the prosperous regions of irrigated agriculture in the Northwest. Several annual circuits have been established between these central areas and the Northwest, along which agricultural labourers migrate most of the year, returning to their home base only a few weeks or months a year. Others have resettled in regions where work is available all year long. In Zamora, Michoacán, for example, the strawberry export agribusiness has brought in 50,000 agricultural labourers, women and men, who settled there in the seventies.

At the same time that labour was being released from agriculture—as small farmers needed more wage labour to perform and large farms began to mechanise—the rate at which employment was being created in industrial centres began to fall. In Mexico City, it decreased from 4.9 % for men, and 5.0 % for women in the fifties, to 3.2 and 3.3 % in the sixties (Contreras 1972). Many reasons account for the decline in the number of jobs being created in manufacturing industry in Mexico, including restricted markets, insufficient reinvestments, high costs of imported technology, repatriation of profits by multinationals, and capital-intensive forms of production.

Migrants thus increasingly entered the service sector, which was responsible for 30.1 % of new jobs in the forties, 33.2 % in the fifties, and 55.5 % in the sixties (Contreras 1972: 408). But it has been the informal sector, that is, non-contractual, low-paid jobs usually in self-employment that has overwhelmingly taken in the rural migrants since the sixties (Arizpe 1975; Lomnitz 1978).

In this section, some of the major indicators of the crisis of small landholdings in Mexico have been mentioned, but these give only a partial picture of rural out-migration. In the end, it is the way in which these major economic pressures are absorbed at the local level that contributes decisively to migration patterns. One must look at their *local effects* as well as at the community organization, the local political structure, and the division of labour within the peasant household. These are the conditions that most clearly explain the different types of migrants who have left and are leaving the Mexican countryside. Now we can ask, given these necessary conditions for migration, what have been the sufficient conditions that have selected certain groups and individuals for migration?

3.5 Social Structure in Rural Communities and Outmigration

The integration of migrant communities in Mexico into the national market economy in the last three decades has entailed a total rendering of economic and social relations: single factors alone cannot account for the changes, or for their consequences. This explains, as I hope will become clear in the analysis that follows, why correlation analysis of variables of village and municipal conditions and out-migration fails fully to explain migration (Stoltman/Ball 1971: 47–56).

"Love recognizes no barriers….penetrates WALLS…", Graffiti of a quote by Maya Angelou on the Mexico–US border fence in Tijuana, 2012. This photo was taken by Lourdes Arizpe and is from her photo collection

An example can help illustrate this problem. A cash-crop has been found in some cases to deter migration, as it both enables peasant families to obtain much-needed cash locally and stabilises seasonal fluctuations of employment.[3] The more dynamic commercial flow that is produced may bring to the villages some of the goods and amenities that make cities so attractive. On the other hand, a cash-crop may encourage migration, when its introduction has the consequence of centralising land and capital. This displaces small landholders and, by fostering capital intensification of cultivation, also makes agricultural labourers redundant. It can also force peasant households to send out migrants in order to get cash needed to plant and harvest the commercial crop.

This example emphasises the extent to which the social and political structure of a town or community will soften or sharpen the negative aspects of integration into the market economy. Thus, a uniformly structured abstract 'rural community' cannot be assumed.

Another example of this kind is road building. One of the more notable achievements of the Mexican government has been the building of hundreds of kilometres of paved and gravel roads that have opened the way for dynamic exchange of migrants and commodities to even the most remote villages. Clearly, a new road will create new migrants, or ease the discomforts of travelling for old migrants, but this in itself need not be negative. That is, it is necessary to assess the optimal point at which a 'healthy' outmigration, which may ease pressures on community resources and move labourers to more productive jobs, turns into a damaging drain on the community, as the most able migrants as well as vital resources start leaving by the road. Such variation in the impact of road building has been reported by Cornelius for Mexico (1976: 23), Richards (1954) and Safa/Du Toit (1976) for Africa, and by Connell and Dasgupta for India and other developing countries (Connell/Dasgupta/Laishley/Lipton 1976).

Against this background, I will summarise the way in which greater integration of peasant communities into the Mexican economy has influenced the rural exodus in the past three decades.

First, the peasant economy in Mexico is not as 'traditional' as some may think. The agrarian reform brought about a considerable resettling of the rural population, making it possible for communities to reconstitute a peasant economy based on self-subsistence production in households closely linked by reciprocity within the community. Two characteristics of the agrarian reform land distribution were to have important repercussions in later outmigration. The first was that in many *ejidos*, that is, the newly formed cluster of allocative lands, the claim to the usufruct of land was granted on a collective basis, with no individual allotment of land plots. The *ejidatarios* could cultivate as large a plot as they were able to care for, provided enough land was available. Since at that time most of them had little agricultural machinery, it happened that those with larger households, especially with a larger number of sons, had an advantage over others in being able to cultivate more land.

---

[3] This, for example, was the case of San Francisco Dotejiare, a community that still derives income from the traditional crop of the *zacatón* root, reported in Arizpe 1978.

Thus, 'human capital,' even in this initial period, was far more reliable in the eyes of the farmers than 'economic capital,' which was extremely scarce in any case.

In highly populated areas, the average allotment of land granted in the *ejido* was sometimes less than 6.5 hectares considered a *minifundio*, that is, a plot of land that does not allow an average farming household to make its livelihood from it. In Santiago Toxi, a community in the state of Mexico, the *ejidatarios* started out with 2.5 hectares of land in 1929; by the seventies, most of the *ejido* plots, and even the private plots of land, were hardly one hectare on average (Arizpe 1978: 98). Many young men wanted to stay in the village and continue to farm, but as the 20-year-old son of Pascual de la Luz put it: "I want to work on the land, yes, but tell me, what am I going to do with a few furrows?"

Second, in the land distribution programs no provision was made for the natural population increase of *ejidatario* households, as Cornelius has also pointed out. In fact, sons of *ejidatarios* frequently do not have legal title to their *ejido* lands, making it extremely difficult for them to set up guarantees for agricultural credit and allowing all sorts of irregularities in the reassignment of land.

Population growth in Mexico has been translated directly into fragmentation of lands (as happened in France, in contrast to England and parts of Germany), due to partible inheritance, the principle drawn up in the Mexican Constitution directly from the Napoleonic Code. In theory, all offspring, including women, have the right to inherit patrimonial land. In fact, women rarely inherit land since, they are expected to marry a husband who will provide land and, indeed, they were legally banned from holding *ejido* rights to land until 1975. Daughters usually receive a few belongings, animals, or some money as inheritance. But all male offspring do have a right to land. The fragmentation of *ejido* lands has already gone through at least two and, in most places, even three or four generations. In Huecorio, Michoacán, the average size of landholdings had shrunk from 5.2 hectares in 1960 to 2.8 hectares by 1976 (Dinerman 1978: 491).

Yet population growth by itself explains neither the crisis of agriculture nor the high rate of rural outmigration. Proof of this is that the amount of arable land per capita in Mexico has not changed since the 1930s[4] (Jenkins 1977: 186). As Connell and his co-authors note for several developing countries, the crude man/land ratio "tells us nothing about the productivity of the soil, the cultivation and, importantly, its distribution among the resident population" (Connell/Dasgupta/Laishley/Lipton 1976: 8).

As noted in the previous section, the concentration of land has continued in Mexico, and the rain-fed farming sector has suffered acutely from demographic pressure on land. Even so, some communities with high population pressure, such as Santiago Toxi, were able to maintain a viable economy based on rain-fed agriculture because they had other options that allowed them to diversify the household economy.

---

[4] Jenkins quotes Rodolfo Stavenhagen and others' extensive work on Mexican agriculture at the Centro de Investigaciones Agrarias during the sixties.

In communities with land scarcity, such as Toxi, the search for alternative sources of income began as early as the forties. Many households could still rely on old-time sources of income: petty trade, crafts, and cottage industries. Some communities in fact depended primarily on such activities. In many regions, villages had specialised in certain crafts which they traded with other villages through weekly markets. Such intricate traditional systems of exchange were highly developed, for example, in the Pátzcuaro region of the state of Michoacán. Similar systems have been amply described in the state of Puebla, Oaxaca, Tlaxcala, Hidalgo, Querétaro, and most other central and southern states.

In other regions, traditional cash-crops for the internal as well as international market still provided a steady inflow of cash. *Ixtle* fibres made from agave and used for rope making were a mainstay in the states of Hidalgo, San Luis Potosi, Guanajuato, and others. The hard, fibrous root of the *zacatón* plant had been exported abroad since last century in regions of the states of Mexico, Michoacán, Oaxaca, and Jalisco.

Crafts had ample demand since they provided the bulk of clothing, housewares, agricultural tools, furniture, horse and oxen apparel, cargo equipment, containers, and toys. So did the cottage industries of brewing, cooking tanning, weaving, and others. The great advantage of such activities, as well as of petty trade, was their adaptability to the seasonal needs of labour by distributing the various activities among the household members by sex and age.

Local sources of wage labour were also available in agriculture during planting and harvesting seasons and in government projects, such as road and dam building programs that were developed in the forties and fifties.

Labour migration was one option among many. Up until the end of the fifties it was mainly seasonal (father and/or sons working in construction work in the cities, or in cash-cropping areas) and temporary (sons and daughters working in the city, males mainly as market porters, handymen, and servants and females as resident servants). Another possibility in regions where recruitment centres had been set up was, of course, migration to the United States through the Bracero Programme.

All these labour options meant that the larger the household, the higher its income, as household members combined agricultural and other activities. Again, it shows that their 'human capital' was clearly an asset. The lowering of the mortality rate, especially infant mortality, enhanced the possibility of survival and accumulation of resources for these households.

If peasant communities were contributing mostly seasonal and temporary migrants, who were the migrants taking up permanent jobs in the cities? It is interesting that during the forties and fifties they were mostly young people with schooling from regional cities and rural towns. They were the first to feel the full force of attraction of the industrial cities (Muñoz/Oliveira/Stern 1977). Part of the reason for their leaving was that the highly centralised pattern of industrial development in Mexico had a stagnating effect on provincial cities and towns. Employment in rural areas was expanding very slowly in non-agricultural jobs, although standards for schooling and social mobility were rising. Thus, young people with secondary or preparatory school education were keen to go to Mexico City where they could live the modern life. There were, of course, a few permanent migrants who did come from peasant communities during this 'pioneer' stage of migration, mostly on an individual basis out of ambition or adventurous impulse, or because of conflicts back home (Balán/Browning/Jelin 1973; Kemper 1977).

During the fifties, as the move to incorporate the peasant sector into the internal market gathered strength, irreversible economic changes began to occur in small rural communities. Previously, the exchange of goods and services within the communities had ensured that any surplus produced would remain within the region. Other socially prescribed mechanisms helped ensure that the surplus of individual households was roughly redistributed among the households of the community. The most important redistributive mechanism was the cargo system, whereby the wealthier members of the community were expected to serve the collective social and religious life of the community by spending on fiestas and ceremonies.

Kinship and residential arrangements also allowed this redistribution. Any household or family group in distress could be bailed out for some time or could simply be reattached or its members redistributed among the more prosperous households. Uxorilocal marriage (where the groom lives with the bride's parents), compound families, and ritual kinship could also achieve this end.

The crucial point here for understanding outmigration is that maintenance of a high degree of interdependence and levelling off of wealth inequalities within the community lessen the economic risks for any given household, thus inhibiting the bankruptcy and landlessness of farming families. This is not to say that inequalities did not exist in such communities. Of course they did, and political power was unevenly distributed due to the *cacicazgos*, but in a self-subsistence economy such social pressures did inhibit the creation of extremes of wealth and poverty.

When these social mechanisms of redistribution are altered and ultimately destroyed, a process begins within the community, heading toward increased social and economic inequality, including landlessness, all of which sets the stage for outmigration. This happens, in my view, not so much as the modernisation of agriculture—or, I would add, the modernisation of social values—but from the monetisation of the economy, as Kate Young has argued for a community in Oaxaca.

The ripple effect of monetisation can be most clearly understood in villages where a cash-crop at some point suddenly offers vast amounts of cash. A case in point is the cultivation of coffee, which increased its price 22-fold from 1938 to 1954 (Young 1982). In one community in the Sierra at Puebla, Zacatipan, the sudden boom in coffee brought about the fragmentation of patrimonial estates owned by several kin-related households, as young sons could now become economically independent of the parental household. They could obtain cash immediately, whereas before, with maize cultivation, agricultural inputs and reciprocal labour could be obtained only by belonging to the kin-based labour groups. As a result, the young neo-local households became entirely dependent on the income of the coffee sales, and with the fall of the price of coffee were at greater risk of having to sell or lose their land and to migrate (Arizpe 1972).

The same process was analysed by Young in Copa Bitoo, in the state of Oaxaca. Young reports that wage labour became prevalent during the peak of the coffee boom. When coffee prices in the international market fell in the mid-fifties, households were unable to revive the traditional *manovuelta* system of labour exchange; they had to revert to unpaid family labour, since they could no longer pay for labourers. But by then many households had had to send sons and daughters outside the community to sell their labour. This seasonal and permanent outmigration further eroded the system of exchange of labour within the community. A large majority of households then had to decrease the amount of maize cultivated to conform with the labour they could rely on within the household (Young 1982: 24–25).

Changes flowing from the monetization of the rural economy were often slower than in these examples but were pervasive in all rural areas of Mexico. The increase in cash needed to pay the production costs of maize was the most crucial development. By 1973 in Santiago Toxi 80 % of the production costs of maize had to be paid for in cash, compared to 30 % a decade earlier. Kin-based labour and plough-sharing groups disappeared, as did the exchange of seed among kin-related groups. Now, because of soil erosion from constant cultivation, fertiliser was indispensable and expensive. Now a plough and oxen had to be rented, as well as mules for other cultivation tasks.

Meanwhile, income from the sale of maize and other agricultural products bought less and less in the market. The farmers perceived that their growing deficit hinged on the price of maize. "Everything goes up (in price), but not maize. Why do you think that is?" asked Raúl Martínez. Another farmer quickly summarised the whole situation: "It is not viable to plant maize anymore because the prices of everything have gone up. Take fertiliser, one year it costs 600 pesos, the next 700 pesos. Meantime, the price of maize had not gone up for fifteen or twenty years. That's why people don't want to plant anymore and they'd rather go to Mexico City to work" (Arizpe 1978).

During this same period, local cottage and crafts industries greatly declined. Manufactured products poured into rural areas; some were more durable than their local counterparts, for example, tin pots instead of pottery; or cheaper, such as factory-woven blankets; or had greater prestige, such as plastic flowers compared to hand-woven decorations. Bottled beer swept away locally brewed beverages such as *pulque*, a formerly very popular alcoholic drink which lost 80 % of its price value, while junk foods and soft drinks replaced women's sales of tortilla savouries, *quelites* (edible plants), and quick-fried foods. Tule rain capes were replaced by plastic sheets, leather sandals by plastic shoes, reed mats by mattresses, *ixtle* and *sisal* ropes by synthetic fibre ropes, *zacatón* brushes by those made also with synthetic fibres—the list can go on and on, region by region. Importantly, the old cash-crops mentioned (*zacatón*, *sisal*, *ixtle*, *candelilla* and others) were no longer in demand in the internal or the international market. Also important, the major income-generating activities of women (weaving, sewing, pottery-making, the sale of food and gathered products, petty trade, etc.) all declined, thus lowering poorer household's income and making it almost impossible for women in female-headed households, or on their own, to make a living in these communities.

The critical economic imbalance created in these communities is made even more evident by the fact that at the same time they were losing traditional sources of cash, their cash needs were soaring. New services had to be purchased: electricity, potable water, transport. And the 'modernising' attitude spread by the urban-educated teachers and the mass media, aggressively and with contempt for the rural way of life, has fostered the consumption of urban prestige goods: fashionable clothes, sumptuous consoles, record players, and so on.

As the national culture became more geared toward the urban, industrial setting, which acquired prestige as the more 'modern' way of life, many young men and women came to see agriculture and life in rural towns and villages as an 'uncivilised' and certainly unsophisticated existence. As one wealthy farmer put it: "[Young people] don't want to work here anymore. The boys come along with wristwatches and brace-lets. Agriculture is dead, there aren't any labourers…You can't find a puny little maid anymore. The girls who go to school get full of ideas." It is not only a question of 'ideas', however, but also of a direct contrast of the rewards—social and economic—offered by rural and urban work. An example is a young migrant´s view of working on the *zacatón* root: "Pulling the root out of the *zacatón* is very hard work. You have to get up at 6 am and keep loosening the plants with the stake. Then you have to take off all the lumps of soil. You get covered with dust; you suffer a lot to get it out."

In the city, one rural migrant explained: "life is much better. There you can go to the movies and you don't have to kill yourself to make a living. Here, in the countryside, you don't have the security of a salary, because in cultivating maize, many times you end up losing, since you don't get back what you put in, in costs of fertiliser, oxen, and the labourers you have to pay. That's why many from the village go there (Mexico City) with their women, and sometimes even take their children. Over there, the woman and the man can work, and the money is certain." Thus is how rural men and women have perceived the changes around them inher-ent in the policies described in the previous section and the way they have reacted to them. It shows that for all their 'traditionalism', rural people will not endure a hopeless, exploitative situation for long. But other factors made it extremely dif-ficult for people in the communities to improve their situation locally.

Once the amassing of goods became the main source of prestige—and of political power—the wealthy in the communities began a pattern of conspicuous consump-tion. But access to credit and business concessions runs through political channels. The political and the economic elite thus reinforced each other by drawing a closer circle of overlapping power and wealth, while the subordination of the judicial sys-tem to the political class gave no standpoint from which to check abuse. Thus was consolidated a rural ruling class which centralised agricultural, financial, commer-cial, and political capital. Against this formidable clique, neither the small landhold-ers, the *ejidatarios*, nor the agricultural labourers have any political leverage.

Significantly, this rural ruling class also has a high incidence of sons and daughters migrating for a better education or to live in more fashionable surround-ings. The same phenomenon has been reported in other developing countries.

In the conditions described, households become increasingly reluctant to fulfil community ritual and social obligations. The maxim of *"que cada quien se rasque con sus propias uñas"* ('let everyone scratch with their own fingernails') is quickly becoming the pervasive value for social behaviour. As a consequence, the services and goods provided by people in the community for social ritual are no longer in demand. Gone are the dancers, the musicians, the costume-makers, the prayer women, the chapel decorators, the makers of ritual ornaments. Gone is the free eating and drinking for the poorer members of the village.

In sum, work opportunities in the village were stripped to the barest bones. Local wage labour provided neither expanded opportunities nor adequate amounts of cash. In Huecorio, Michoacán, the day wage in agriculture was 6 or 7 pesos (US 48¢ or 56¢) in 1962; by 1976, that is, fourteen years later, when agricultural revenues in the rain-fed farms had plunged and cash needs had gone to the stratosphere, the daily wage in agriculture had soared to 8 pesos (US 64¢) a day! (Dinerman 1978: 491).

But the discrepancy with urban wages and salaries had the greatest effect on migration, particularly in regions within a 500-kilometre radius of the cities. In regions outlying Mexico City (states of Hidalgo, Tlaxcala, Puebla, Morelos, México, and Querétaro) the official minimum salary was 18–21 pesos (US $1.44–$1.68) per day, in 1972; the wage actually paid in these areas was 8–10 pesos (US 0.64–80¢) for women and 10–15 pesos (US 80¢ to $1.20) for men. Meanwhile, the minimum wage in Mexico City—only two hours away by bus—was 41 pesos (US $3.28); a construction worker could easily make 25 pesos (US $2) a day, and a street vendor at least 20 pesos (US $1.60) a day (Arizpe 1975).

Needless to say, the discrepancy with wages across the border has been even more blatant. In 1976, Cornelius reported that local agricultural wages in Jalisco were 25–30 pesos (US $2–$2.80) per day, and their equivalent in the United States, $2.50–$3.00 an hour. In US factories the wages were as high as US $4.00 and $5.00 an hour (Cornelius 1976: 23).

The differential between agricultural income levels and wage levels in urban areas and in the United States, in my view, was the crucial pull factor that accelerated outmigration and determined the farming household's strategy. As their deficit in maize cultivation grew, *minifundio* households turned to a multiplication and diversification of wage earnings. Since circumstances did not allow economic investments in a dwindling agricultural enterprise, households directed their meagre surplus to investment in 'human capital'.

This does not mean that parents consciously try to have more children. It means only that they have encountered no reasons to limit their fertility now that more children survive. Indeed, to the contrary, they have found that more children enhance the household's chances of surviving as an agricultural enterprise.

Table 3.1 shows the internal labour composition of households in a sample of 150 households in the villages of Toxi and Dotejiare (Arizpe 1982). In Toxi, local sources of income have disappeared, and land plots are so small that they provide work for only one full-time agricultural worker. As can be seen in the table, at least a third of the household workers are in migratory wage labour all along the developmental cycle of the household. Many unpaid domestic labourers in Toxi, all of them women, also work locally in non-resident, paid, domestic service.

The contrast with Dotejiare is very marked: there, land plots are slightly larger, and two local sources of cash still bring money periodically into the households. These are the brewing and selling of *pulque* and the sale of the root of *zacatón* bush,

Table 3.1   Percentage of workers in households, according to type of work performed and to the stages of the domestic cycle, in two Mexican rural villages

Domestic cycle[a] (mother's age)		Agriculture	Unpaid domestic work	Local wage labour	Migratory wage labour
18–25	Toxi	0	42	25	33
	Dotejiare	50	45	5	0
26–30	Toxi	7	47	3	43
	Dotejiare	51	38	4	7
31–38	Toxi	18	39	8	35
	Dotejiare	61	34	0	5
39–44	Toxi	27	30	9	34
	Dotejiare	61	26	0	13
45–50	Toxi	22	30	7	41
	Dotejiare	43	32	6	18
50 and older	Toxi	35	19	10	35
	Dotejiare	54	32	4	10

[a] Insufficient information on four cases (2.8 %) in each village. Data was estimated by dividing the total number of household workers in specific types of work, for each household, which was then averaged among the households in each stage of the domestic cycle

a communal cash-crop. Consequently, migratory wage labour does not occupy more than one-fifth of the workers in the household at any time during the domestic cycle.

The vital role of migration for the survival of Toxi households and the pattern of what I have called *relay migration* become even clearer in Table 3.2.

There is no clear pattern of outmigration in Dotejiare. In Toxi, on the contrary, it follows a stepwise progression of the first son/daughter at each particular developmental stage taking up migratory wage labour, substituting for the father or the eldest siblings as they separate from the household. Remittances from migrants are used primarily to buy fertiliser and to pay for tractor or plough rental, and also for food.

This pattern of relay migration may apply only for some regions, but it helps explain why the bulk of seasonal and temporary migrants are either small land-holders or sons and daughters of such households. It also warns against the danger of pointing at the 'population explosion' as the culprit of rural outmigration, when, in fact, this may very well be a major life-line for impoverished farmers.

3.6 The Rural Exodus and the Selectivity of US-bound Migrants

Data from the surveys conducted on Mexican migration to the United States, set against the background of the rural exodus analysed in the previous pages, clearly show that those migrating across the border are specific types of migrants.[5]

---

[5] I base this analysis on data from Wayne Cornelius and Ina Dinerman's fieldwork, as well as on the following surveys: Jorge Bustamante (1977) and Francisco Alba (1978).

Table 3.2 Percentage of household members who migrate, according the stages of the domestic cycle

Domestic cycle[a] (mother's age)		Father	First son/ daughter[b]	Second son/ daughter	Third son/ daughter	Others[c]
18–25	Toxi	84	–	–	–	16
	Dotejiare	–	–	–	–	–
26–30	Toxi	100	–	–	–	–
	Dotejiare	–	80	–	–	20
31–38	Toxi	66	29	5	–	–
	Dotejiare	33	33	33	–	–
39–44	Toxi	9	43	43	5	–
	Dotejiare	–	84	16	–	–
45–50	Toxi	8	55	21	11	5
	Dotejiare	–	70	–	10	20
50 and older	Toxi	4	61	23	12	–
	Dotejiare	17	50	17	16	–

[a] Insufficient information on four cases (2.8 %) in each village
[b] This refers to *resident* offspring age order
[c] Includes father's brothers, adopted offspring, sons- and daughters-in-law, and nephews and nieces

In other words, just as migrants are not a tandem sample of rural inhabitants in Mexico, so US-bound Mexican migrants are not a random sample of people leaving the Mexican countryside.

Several community studies in Mexico have shown that many of the poorest, landless people generally tend to stay on in rural areas or to migrate to other rural destinations within Mexico. In Oaxaca, Young (1982) did find that the poorest migrated, first expelling their children, then as whole households, but practically all went only as far as Mexico City or Oaxaca City. Dinerman, from fieldwork in Michoacán explains it in this way: "The landless, those without the resources to build and maintain wide social networks linking them to other households, those who are not prominent in community affairs and thus lack economic allies, do not sponsor migrants" (Dinerman 1978: 498). My own research in the states of Puebla, Mexico, and Michoacán supports this conclusion. The same deterrents seem to apply even more strongly for migrants going to the United States. As Cornelius points out, those at the bottom cannot afford to pay the travel fare or the cost of the '*coyote*' to get across to the United States (Cornelius 1976).

It does seem to be the case that the landless migrate more from villages that have established routes and communities in the United States, as Cornelius reports in his study in Jalisco. This could mean that the village already provides the social networks whose absence, in Dinerman's view, otherwise deters migration of the landless.

Few migrants from middle- and upper-income groups leave Mexico because they seek social and economic mobility and most already have favourable kinship or social contacts in the cities. For different reasons, widowed, divorced, or unmarried women also tend to settle in Mexican cities, unless the village migratory network makes it possible for them to travel across the border. Elderly people leaving the villages go to live with the offspring who have attained the highest standard of living, usually in Mexican cities; few venture to the United States because of the barriers to their crossing and their preference for the more familiar surroundings.

Who are the US-bound migrants, then, and why do they leave? The motive most frequently expressed for migrating to the United States is higher wages. The importance of the effect of wage differentials has been established. On the basis of a longitudinal analysis of Mexican migration to the United States, Jenkins concludes that "it is fluctuations in wage differentials, created largely by changes in Mexican wages that shape migration (to the United States)" (Jenkins 1977: 184).

This is in line with Todaro's model of migration where "migrants are induced to move even though there is a high probability of unemployment, as long as the product of the wage and the probability of employment exceeds the rural wage by sufficient margin to offset the cost of moving[6] (Todaro 1976)". This seems to be borne out in the case of rural outmigration in Mexico. It may even happen, as data from recent fieldwork in Zamora, a booming industrial town in Michoacán, have shown, that young men to whom factory employment is available still prefer to migrate to temporary jobs in the United States.

This does not mean that the flow of Mexican migrants to the United States is made up primarily of workers who chose between two wage packages. Jenkins adds that "overall changes in agricultural productivity and capital investment, not labour conditions, have the major impact on both illegal and bracero migration" (Jenkins 1977: 184). Significantly, he found that the official bracero program enlisted more wage labourers while undocumented migration takes in more landholders.

Thus, the evidence strongly indicates that the bulk of Mexican migrants to *the United States comes from the crisis of small-landholding, rain-fed agriculture*— the *minifundios*, both private and belonging to *ejidos*—that has been described in this chapter.

Why should farmers be more interested than landless labourers in seeking temporary employment in the United States? Some reasons are, probably, that the security of a land base makes these small-holder migrants more willing to risk—and lose—several years of uncertain and seasonal work in the United States, while labourers (male and female) need to find more stable conditions for making a living and, therefore, might prefer Mexican cities and other rural areas. Also,

---

[6] Todaro's model is useful in explaining why migrants leave, but he is open to criticism for not incorporating into his model the informal or traditional urban sector that takes in a majority of migrants in many cities in developing countries (Todaro 1976).

labourers living on a day-to-day basis find it more difficult to scrape together the money needed to travel and to cross the border. Finally, it is very probable that the social networks and information channels necessary for successful migration to the United States can be maintained in a stable way only by land-based migrants.

But although Mexico's agricultural crisis makes available migrants for the US job market, not all of those needing jobs or better prospects end up in the United States. To understand why, we must look at the selectivity of these migrants. Among migrants of small-landholding families we find then (a) the father who has no one else to substitute for him in migrating; his is a seasonal or temporary migration and rarely does he remain in the United States permanently; (b) eldest sons who are sending back remittances and who are expected to fend for themselves in the world since they will not be inheriting; they go temporarily, but tend to stay if conditions allow them to and if they like it; (c) younger sons who will be sending remittances, but who are certain to inherit some land; they most probably return, unless they find very good fortune in the United States; (d) daughters who are also to send remittances but go only if they can travel with the father, a brother or other close related kin, and (e) collateral members that the farming family can no longer afford to employ or to keep; these relatives (e.g., nephews, nieces, cousins) are especially prone to migrate.

3.7   Conclusions

The intensity with which rural outmigration has occurred in Mexico in the last three decades has resulted from the simultaneous effects of, on the one hand, the demand for labour in the expanding industrial and commercial centres in Mexico and in the US agriculture and low-grade urban employment, and on the other, the gradual undermining of the Mexican economy based on small-holding, rain-fed agriculture. Population growth in urban areas in Mexico, far from being at the root of the problem, is itself the result of desperate need in poor households for wage earners. Because opportunities for wage labour are not available in rural areas in the numbers needed, many household members have to become migrants.

The economic mechanisms behind the exodus—the monetisation of the rural economy, the fluctuations and decline of agricultural prices, the destruction of rural industries and occupations—are not unlike those that expelled millions of European migrants overseas, during the mid-nineteenth and early twentieth centuries. Massive rural outmigration, then, is not a new phenomenon in Western industrial capitalism and is by no means an unfamiliar one for the United States.

But the speed at which rural migration has occurred since the fifties in Mexico, as well as in many developing countries, has been undoubtedly influenced by development policies that encouraged industrialisation and urbanisation at the

"Wall of Shame", Graffiti on the Mexico–US border fence, Tijuana, 2012. This photo was taken by Lourdes Arizpe and is from the author's photo collection

expense of agriculture. Many believe that the present rural crisis in these countries can be traced back to the uncritical acceptance of such policies on the part of national governments and international finance institutions.

In the Mexican case, it is clear that the break-up of small-landholding agriculture created the conditions for massive rural outmigration. Specific actions, such as the holding down of the price of maize between 1957 and 1973, only accelerated a process of economic change that was implacably under way. This is not only that this process is irreversible or that it cannot be modified. Already some changes are in view, notably massive government programs in support of rain-fed agriculture. The rate of population growth has also begun to decline; this will not solve the crisis but will help alleviate some of its symptoms. For example, Warman (1978) reports that some farming households in the state of Morelos are reverting back to the extended family arrangement because they have been able to intensify their agricultural production in vegetables. Another example is the new demand for handicrafts among the urban middle-class, which is enabling rural artisans to make a living once again.

Even given the release of agricultural labour, though, migration will occur only if strong factors of attraction arise elsewhere, and these determine the rate of outmigration and the destination of migrants. From the forties to the mid-sixties, growing employment in irrigated agriculture and in the urban industrial and service sectors in Mexico incorporated several million rural migrants. Since the beginning of the sixties, though, the modern sector has been unable to offer employment at the rate required, in the same way as Western European early industries were unable to retain all rural migrants, who were thus forced to migrate overseas. It is important to stress, then, that in the Mexican case the fact that new industries and new agricultural enterprises tend not to be labour-intensive has directly affected migratory flows. In other words, given the availability of migrants in rural communities, the conditions in the place of destination, such as a potential job or family and village networks, determine who will go where.

It is not surprising, therefore, that migration to the United States, both official and undocumented, attracts only certain kinds of migrants. Compared to the general outflow of rural migrants, the group going to the United States contains a lower proportion of the poor and landless, fewer women, fewer elderly people, and fewer young men and women from middle- and higher income rural households. The bulk of migrants to the United States are adult men attracted in some way to small-landholding units. And, which is important, not all of them are unemployed, which means that the pull factors override the push factors in encouraging part of the Mexican migration to the United States.

Thus, since the United States is not receiving all the potential migrants let loose by the rural decline in small-holding farming areas, there is reason to conclude that it is not bearing the brunt of the rural crisis in Mexico. Rather, US employers are benefitting from the crisis by reaping the best gains for their labour market. The case for this is strengthened by noting that these able-bodied, enterprising migrants have been fed, clothed, and educated by the already weakened Mexican rural communities. While it is clear that, for the time being, the Mexican economy, both rural and urban, is unable to accommodate all rural migrants productively, it goes without saying that it is unacceptable that Mexican rural households should become the providers of nursery, social security, and unemployment services for workers in US jobs—workers who can be sent back in periods of economic recession, overburdening an already burdened Mexican rural economy.

Therefore, if it can be ensured that Mexican migrants do not damage the position of US workers, the movement of migrants between Mexico and the United States can constitute a partnership that benefits both sides. But this means that a policy scheme must be sought that gives proper acknowledgement to both partners' needs and benefits.

As to future trends, looking into the crystal ball—with the evidence and figures close by—it would seem that the pattern and rate of this migratory flow will not vary greatly in the years to come. It will not decrease as long as there is demand for such migrant labour in the United States, a fact quite independent of conditions in rural areas in Mexico. It will not intensify if the Mexican government is successful in helping small-holding agriculture, as it has pledged to do through multiple rural development programs. Nevertheless, the massive pouring of financial resources into such communities will take many years to rebuild community social and economic organisation in a way that will begin to have an effect on outmigration. It is clear that what is needed is not only an economic policy, but a comprehensive social policy toward rural agricultural areas.

Under a broader, long-term perspective, it becomes clear that Mexican migration to the United States, involving the lives of thousands of men and women seeking better opportunities, is only part of a more complex relationship between the two countries. The way this relationship evolves overall will set the conditions for solving mutual problems in the future on the basis of reciprocal cooperation.

References

Alba, Francisco, 1978: "Mexico's International Migration as a Manifestation of Its Development Pattern", in: *International Migration Review*, 12, 4, (Winter): 502–513.

Arizpe, Lourdes, 1978: *Migración, etnicismo y cambio económico* (Mexico City: El Colegio de México).

Arizpe, Lourdes, 1972: *Parentesco y Economía en una Sociedad Nahua* (Mexico City: Instituto Nacional Indigenista).

Arizpe, Lourdes, 1975: *Indígenas en la Ciudad: El Caso de las 'Marías'* (Mexico City: Sep-Setentas).

Arizpe, Lourdes 1982: "Relay Migration and the Survival of the Peasant Household", in: Balán, Jorge (Ed.): *Why People Move: Internal Migration and Development* (Paris: UNESCO): 187–210.

Balán, Jorge; Browning, Harley; Jelin, Elizabeth, 1973: *Men in a Developing Society: Geographical and Social Mobility in Monterrey, Mexico* (Austin: University of Texas Press).

Bancomext (Banco Nacional de Comercio Exterior), 1976: 1975: "*Facts, Figures, Trends Mexico 1976*" (Mexico City: Bancomext).

Bustamante, Jorge, 1977: "Undocumented Migration from Mexico: A Research Report" in: *International Migration Review* 11, 2: 149–178.

Butterworth, Douglas, 1971: "Migración Rural-Urbana en América Latina: El Estado de Nuestro Conocimiento", in: *América Indígena*, 31, 1: 52–85.

Connell, John; Dasgupta, Biplab; Laishley, Roy; Lipton, Michael, 1976: *Migration from Rural Areas: The Evidence from Village Studies* (Oxford: Oxford University Press).

Contreras Suárez, Enrique, 1972: "Migración Interna y Oportunidades de Empleo en la Ciudad de México", in: Martínez Ríos, Jorge (Ed.): *El Perfil de México en 1980* (Mexico City: Instituto de Investigaciones Sociales- Universidad Nacional Autónoma de México): 359–418.

Cornelius, Wayne A., 1976: "Out-migration from Rural Mexican Communities", in: *The Dynamics of Migration: International Migration*, Vol. (Washington DC: Smithsonian Institution): 1–40.

Dinerman, Ina R., 1978: "Patterns of Adaptation among Households of US–bound migrants from Michoacán, México", in: *International Migration Review*, 12, 4: 485–501.

García y Griego, Manuel, 1983: "The Importation of Mexican Contract Laborers to the United States, 1942–1964", in: Brown, Peter G.; Shue, Henry (Eds.): *The Border that Joins: Mexican Migrants and US Responsibility* (New Jersey: Totowa): 49–98.

George, Susan, 1977: *How the Other Half Dies* (Totowa, NJ: Allanheld, Osmun & Co).

Gómez Oliver, Luis, 1978: "Crisis Agrícola. Crisis de los Campesino", in: *Comercio Exterior*, 28, 6 (June): 714–727.

Hewitt de Alcántara, Cynthia, 1979: *La Modernización de la Agricultura Mexicana* (Mexico City: Siglo XXI Editores).

Jenkins, J. Craig, 1977: "Push-Pull in Recent Mexican Migration to the U.S." in: *International Migration Review*, 11, 2: 178–189.

Kemper, Robert V., 1977: *Migration and Adaptation: Tzintzuntzan Peasants in Mexico City* (Beverly Hills, California: Sage Publications).

Lomnitz, Larissa, 1978: *Como Sobreviven los Marginados* (Mexico City: Siglo XXI Editores).

Muñoz, Humberto; de Oliveira, Orlandina; Stern, Claudio, 1977: *Migración y Desigualdad Social* (Mexico City: El Colegio de México-UNAM).

Paré, Luisa, 1977: *El Proletariado Agrícola de México* (Mexico City: Siglo XXI Editores).

Reyes Osorio, Sergio (ed.), 1974: *Estructura Agraria y Desarrollo Agrícola en México* (Mexico City: Fondo de Cultura Económica).

Richards, Audrey I. (Ed.), 1954: *Economic Development and Tribal Change* (Cambridge: W. Hefter & Sons).

Safa, Helen I.; Du Toit, Brian, 1976: *Migration and Development* (The Hague: Mouton Publishers).

Schumacher, August, 1981: *Agricultural Development and Rural Employment: A Mexican Dilemma* (San Diego: Centre for Mexican-US Studies, University of California).

Solís, Leopoldo, 1967: "Hacia un Análisis General a Largo Plazo del Desarrollo Económico de México", in: *Demografía y Economía*, El Colegio de México, 1, 1 (April–June): 23–70.

Solís, Leopoldo, 1970: *La realidad económica mexicana: retrovisión y perspectivas* (Mexico City: Siglo XXI).

Stoltman, Joseph P.; Ball, John M., 1971: "Migration and the Local Economic Factor in Rural Mexico", in: *Human Organisation*, 30, 1: 47–56.

Thomas, Brinley, 1996: *International Migration and Economic Development* (Paris: UNESCO).

Todaro, Michael P., 1976: *Internal Migration in Developing Countries* (Geneva: ILO).

Warman, Arturo, 1978: "*Política Agraria a Política Agricola*" in Comercio Exterior 28, no. 6 (June): 681–687.

Young. Kate, 1982: "The Creation of a Relative Surplus Population: A case study from Mexico", in: Benería, Lourdes (Ed.): *Women and Development. The Sexual Division of Labor in Rural Societies* (New York: Praeger): 149–178.

Issues in culture and development

Cultural changes
in women's life-stages

UNESCO/Society for International Development

Report of the Seminar on 'Cultural Changes in Women's Life Stages' organized by Lourdes Arizpe at UNESCO and presented to the UN Conference on Women, Beijing 1995

Chapter 4
Migration, Gender and Global Crises

4.1 Women Migrants' Uneven Gains and the Impact of the Recession

The impact of the recession, given the gendered labour market, especially among migrants, has widened the gap between the gains made by women as they live and work in other countries and those made by male migrants.[1] Evidence from previous crises suggests that women and women immigrants suffer disproportionately during global crises, having fewer and less effective buffers to cope with economic hardships. Failure to take such groups into account in government strategies in many cases is aggravating their working and living conditions; it widens inequalities and it reduces the acceptance of migrants among receiving communities.

The diversity of women's migratory trajectories, types of labour insertion, income, and motivations makes it difficult to generalize as to the recent impact of the crisis and of policy responses by governments (Kofman/Raghuram 2009). While the number of women in cross-border migrant flows has increased in some regions in the last few years (UN 2005; UNFPA 2006), it is not clear whether this has helped alleviate poverty and how it has benefitted different groups of women (De Haan/Shahin 2009).

The 'feminization' of international migration is linked to the growth of the care economy in developed countries. Women's joblessness is now rising at a higher rate due to public sector budget cuts in education, health and social services. They now have to pay for the care of children, the elderly and the sick in their own homes. Such jobs are relatively more stable, but with the global crises, as more migrant men than women are losing their jobs in construction, agriculture, gardening and retailing, a greater burden is placed on women migrants' shoulders to send remittances home.

In terms of destination, women's migratory flows have increased to Europe, the United States and Asia (Fix/Papademetriou/Batalova/Terrazas/Yi-Ying Lin/Mittelstadt 2009). An estimated 46.2 million (51 %) of all migrants have OECD

[1] This text was presented at the meeting of the Committee for Development Policy of the Economic and Social Council (ECOSOC) of the United Nations, of which Lourdes Arizpe is a member, in March 2009 and is unpublished.

Table 4.1 Women's cross-border migration 1990–2009

	1990	1995	2000	2005	2010
Total migrants	155,518,065	165,968,778	178,498,563	195,245,404	213,943,812
Male migrants	79,132,432	84,207,529	90,242,214	99,171,119	109,148,850
Female migrants	76,385,633	81,761,249	88,256,349	96,074,285	104,794,962
Female migrants as percentage	49.1	49.3	49.4	49.2	49.0
		1990–1995	1995–2000	2000–2005	2005–2010
Net increase overall		10,450,713	12,529,785	16,746,841	18,698,408
Net increase—males		5,075,097	6,034,685	8,928,905	9,977,731
Net increase—females		5,375,616	6,495,100	7,817,936	8,720,677
Females as percentage of growth		51.4	51.8	46.7	46.6

Source IOM (2009)

destinations and 8.7 million (40 %) go to high-income non-OECD countries (Ratha/Shaw 2007). In 2005, women represented 53.4 % of migrants in Europe, 50.4 % in North America and 45.5 % in the South (Martin/Martin/Cross 2007). While the US accounts for one in twenty of the world's residents, it is home to one in five of the world's migrants (IOM 2009).

The total number of migrant women, however, has remained stable in the last three decades, while female cross-border migration peaked in the nineties. Regional variations show differential patterns: female migration growth reached 66.5 % in East Asia, especially the Philippines, and 60.7 % in Central Asia, as shown in Table 4.1.

Women migrants tend to be heavily concentrated in labour-intensive export industries—which have been strongly hit by the recession—and in low-skilled service industries and domestic care work—for which no current data are available. Women migrants have tended to constitute a layer of double buffer workers as the export market dynamics have generated a flexible and mobile workforce of casuals, temporary workers, contract workers and homeworkers. Women in this mobile, temporarily employed workforce suffer great hardship as they are highly vulnerable to being pulled into the low-skilled entertainment industry, prostitution and the drug trade. A horrifying example is the flocking of young women to the '*maquiladora*' industries of the Mexican border; they are now caught in a jobless, lawless and violent environment in which more than 600 women have been killed, and for which the new term 'feminicides' has been coined.

Internationally, women lead in the brain drain, as the difference in the number of men and women migrants with tertiary education is 3.2 % in Latin America and the Caribbean, 8 % in Oceania and 10 % in Africa (UNIFEM 2008/2009). No data are available on their present situation, although in previous crises evidence suggests they are the last to be hired and the first to be fired.

Deportations of migrants, as well as curbs on the arrivals of migrants, have increased in many developed countries during the recession. For example, 546,039 Mexicans were repatriated between September 2008 and July 2009.[2] Among them

---

[2] President's Third State of the Union Report, 2009. Available at: http://www.inaugural.senate.gov/swearing-in/addresses (1 November 2013).

were 19,000 minors, including 4,317 girls, all of them travelling without any adults accompanying them according to the National Migration Institute.[3]

4.2 Remittances

In some places, remittances are increasing in importance as poverty strategies fail and unemployment rises, making some families more dependent on remittances than ever before. Remittances, however, at least at the outset of recession, have proved to be less volatile sources of income than either exports or foreign investment, as in the case of the Philippines. According to the Institute for Migration Studies, though remittances have dropped globally amid the downturn, some regions are experiencing remittance increases or are holding steady.

4.3 Return Migrants

Some evidence indicates that temporary migrants, among them young women who work as domestic careworkers, waitresses, paramedics, or shop attendants are resisting returning to their countries. As the Migration Policy Institute study reports "…people's decisions whether to return home are predicated on what is happening in the sending-country economies to an even greater degree than changes in the receiving country economy" (Fix/Papademetriou/Batalova/Terrazas/Yi-Ying Lin/Mittelstadt 2009: 3).

References

De Haan, Arjan; Shahin, Yaqub, 2009: *Migration and Poverty: Linkages, Knowledge Gaps and Policy Implications* (Geneva: United Nations Research Institute for Social Development).

Fix, Michael; Papademetriou, Demetrios G.; Batalova, Jeanne; Terrazas, Aaron; Yi-Ying Lin, Serena; Mittelstadt, Michelle, 2009: *Migration and the Global Recession* (Washington: Migration Policy Institute).

IOM (International Organization for Migration), 2009: *Global Estimates and Trends* (Geneva: IOM).

Kofman, Eleonore; Raghuram, Parvati, 2009: *The Implications of Migration for Gender and Care Regimes in the South* (Geneva: United Nations Research Institute for Social Development).

Martin, Philip; Martin, Susan; Cross, Sarah, 2007: "High-level Dialogue on Migration and Development", in: *International Migration*, 45,1 (March): 7–25.

Ratha, Dilip; Shaw, William, 2007: *South-South Migration and Remittances* (Washington DC: The World Bank).

UN Population Division, 2005: *Proceedings of the United Nations Expert Group Meeting on Social and Economic Implications of Changing Population Age Structures* (New York: UN Population Division).

UNFPA (United Nations Populations Fund), 2006: *UNFPA State of the World Population 2006: A Passage to Hope. Women and International Migration* (New York: UNFPA).

UNIFEM (United Nations Development fund for Women), 2008/2009: *Who Answers to Whom? Gender & Accountability* (New York: UNIFEM).

---

[3] National Migration Institute (INM) reported in *La Jornada*, 11 October 2009: 9.

Participating in the Symposium on Research on Women, Wenner-Gren Foundation, Burg Wartenstein, Austria in 1985. Among its participants are Helen Safa, Eleanor Leacock, Lourdes Benería, Achola Pala Verena Stolke and Kate Young. This photo is from the author's photo collection

Part II
Women and Development

Plenary Session of the First Mexico-Central American Symposium on Women Studies, of which Lourdes Arizpe was Secretary-General, which took place in Mexico City in 1977. This photo is from the author's photo collection

Chapter 5
Women Moving Across Boundaries: Movements and Migrations

5.1 Introduction

In the last decades, women have moved across many boundaries: in labour participation and professional work, in political commitment and elected office, in scientific research and education, in cross-border flows, and in daring to question why theologies and churches exclude them from hierarchy and power.[1] The discourse and narratives of women's hopes have been greatly enriched by a myriad experiences in different arenas and by soaring demands not only to be heard but to have authority and influence in decision-making. The diversity and richness of this movement has intertwined with neoliberalism with very mixed results. While neoliberal policies have opened up economic opportunities that diverse groups of women, both rich and poor, have benefited from, such policies have failed to inspire the profound changes needed to balance men's and women's roles in paid and unpaid work, in gaining freedom and entrepreneurship, in providing care and sociability, and in receiving respect and recognition. In all these arenas, what women receive and what they give decisively influences the well-being of the whole of society. When social injustice is added to gender inequality, development is held back and a sustainable future is not possible.

Democratic societies are built on the core values that are also emphasized in feminism: self-reliance, participation, and freedom from violence. In some societies that still have atavistic values of non-recognition and disrespect for women, the struggle for democracy has now strengthened fault lines of resistance to change as far as the role and status of women are concerned. In spite of this, these core values of democracy and feminism have been adopted by active women's movements all over the world, with each group deciding at what speed, at what rhythm, and with which priorities they will be moving across hitherto forbidden boundaries. For some women, human rights and dignity are all that is needed; others emphasize decent work and political influence, as well as selfhood and autonomy. As women's movements broaden the

---

[1] This chapter was presented and discussed at the Committee for Development Policy of the UN Economic and Development Council in New York, 28 April 2009 and is unpublished.

L. Arizpe, *Migration, Women and Social Development*,
SpringerBriefs on Pioneers in Science and Practice 11,
DOI: 10.1007/978-3-319-06572-4_5, © The Author(s) 2014

arenas and issues that they deal with, all the above, in terms of social justice, freedom from war and conflicts, and development, are claims that are made for everyone, for women and men in all countries. As Harcourt (2009: 133) in her editorial for the journal *Development* remarked with reference to the 2008 Forum on the 'Power of Movements' of the *Association of Women's Rights and Development* (AWID) 'Women and Development Meeting' in South Africa: "As ever, the Forum was highly professional, filled to the brim and driven by numerous agendas. Attending an AWID Forum no one could say there is no global women's movement. Nor could one say there is a homogeneous women's movement. Nor that it does not attract the young" (on new activisms, see also Harris 2008). Harcourt (2009) concluded by saying that "… the women's movement is diverse, but definitely… it is global".

5.2 Diversity of Women's Movements

That the global expansion of women's movements reflects great diversity should come as no surprise. Yet there are striking commonalities. A 2008 study by the *United Nations Research Institute for Social Development* (UNRISD) showed that there are many differences and similarities in the way the genders carry out care and non-care work in countries from several different regions, yet, in all countries, the mean time spent on unpaid care work by women is more than twice that for

With Egyptian female students at the Library of Alexandria, Egypt 2010. This photo is from the author's photo collection

men (Budlender 2008: v). It also showed that, when data in the *System of National Accounts* (SNA) and unpaid care work are combined, women are found to do more work than men in all countries (Budlender 2008: 45).

Social science studies have generally shown that diversification is one of the indicators that a political and social movement has expanded beyond small groups to influence the whole of society. Nancy Fraser locates the thrust of the women's movement on the left, as it focused initially in problems of distribution and was drawn, in a second stage, to one of 'identity politics' (Fraser 2004). She argues that the shift from a state-organized variant of capitalism to neoliberalism resignified feminist ideals within the framework of a 'new spirit of capitalism', one that created a new 'connexionist' project for capitalism (as Luc Boltanski and Eve Chiapello have proposed), decreasing traditional hierarchies to give way to horizontal and flexible relationships, thereby liberating individual creativity (Fraser 2009: 109). She is right in saying that a new form of feminist activism—transnational, multiscalar, post-Westphalian—has emerged, yet the impact of such a movement has to be understood in developing countries within the framework of the new inequalities that have brought about high unemployment and the expansion of illicit activities.

Women's movements have been enriched and broadened in the new century: to the traditional demand for political participation and economic autonomy, new activisms have been added. More women of all generations are now present in human rights initiatives, in civil society, and in voluntary associations from the local to the international level. They are very active, as noted in the above-mentioned AWID 'Women and Development Meeting', in struggles against religious fundamentalisms, in sexual and reproductive rights, in communication strategies, in union organizing, and in fighting against gender-based violence (Harcourt 2009).

A trend to be monitored closely is how 'identity politics' is pushing feminism into a very narrow lane. As in other feminist meetings, Harcourt (2009) reports that identity, difference and sexuality issues were in fact the primary issues at the AWID forum, while only a few women attended workshops to discuss hunger, famine, natural disasters, migration, and the global financial crisis. At a time when channels of organizing for public action are disappearing, many women are being driven to take refuge in their private world of psychological and emotional changes. That such a rapidly expanding movement has created a backlash is, as Chun/Boyd/Lessard (2007) argue, an overly simplistic and pre-Foucauldian conceptualization, based on a view of power as omnipotent and one-dimensional, centred on patriarchy, the state or civil society institutions. At the same time, while postmodernism and deconstructivism have opened exciting new perspectives on women's identity and expression in the arts, Scott (1995) warns that they give "an impulse that makes feminist theory impossible".

That such a diverse global movement has achieved more in 'symbolic politics' than in concrete achievements, as Naila Kabeer (personal communication) has observed, is perhaps true. Yet an interesting new development as the economic recession deepens is the opinion expressed ever more often that if more women were in power, the sad state of crisis, warfare, and violence prevalent at present in the world would perhaps be transformed towards a more sustainable way of living.

Interestingly, more women seem to be participating now as militants in conflicts or as military combatants. It is important to note that, at one time, women formed the majority of militants in the Zapatista movement in Chiapas, Mexico, as also seems to have been the case in the guerrilla movement in Guatemala. However, a distinction must be made between two very different kinds of militants: on the one hand, young women who see their path to higher education and autonomy thwarted by traditional male or religious gatekeepers or by the lack of jobs, and so become militants in social and political movements; these women are fighting for change. On the other hand, young women who give up their lives out of religious zeal, such as the young women who have blown themselves up with bombs as part of Islamist terrorist tactics in the Middle East; these women are fighting to stop change. A third group of women are not militants but become involved in violent and frequently criminal activities, such as the Mexican beauty queen Miss Sinaloa; these women may find themselves in the midst of drug trafficking or criminal activities and end up taking this opportunity to gain access to money and power. Do we know why they have all acted in this way? No, since there have been very few studies of the increasingly violent militancy of women.

If most of the gains of the feminist movement are in 'symbolic politics', is some of this active militancy of women an attempt to turn such symbolic politics into real advances? This would be the case, in my view, with the young women involved in the Zapatista army. But not in the case of women retrenching to subordinate positions, or getting involved in violent criminal activities. There is some evidence that, just as happens with men, if doors are closed to women's active participation in mainstream development, women will slip into extremism and violent choices.

This points to a fact that needs to be highlighted: women are no longer passive onlookers at processes that are changing their lives. Nevertheless, their participation is uneven and leads to very different and sometimes unexpected outcomes. How women may gain greater influence through this participation is captured in the title *Developing Power*, the 2004 book edited by Arvon Fraser and Irene Tinker.

It is this realization, I think, that is coming to the fore as the Beijing plus 15 Conference approaches. The Beijing plus 15 review of progress, due in 2010, will require detailed analysis of development policy and programmes, with a call for new, bold perspectives that should influence initiatives for international policy reforms.

5.3 Gendered Impacts of Development Policies

In the last 20 years economic growth policies have given genders uneven and unfair choices, to which must be added that the present economic crisis is already having a gendered effect that may push many women back into poverty or is placing an added burden of unpaid work on them to compensate for the fall in jobs and incomes. As the UN Under Secretary-General for Economic and Social Affairs, Sha Zukang, recently stated "…historically, economic recessions have placed a disproportionate burden on women" (Deen 2009: 3).

Using 'patriarchy' as an analytical concept, Moghadam (1996: 19–20) questions whether development and industrialization have increased or decreased gender inequality. Partly basing her analysis on the Human Development Index, she shows that positive advances have been made in women's life expectancy, literacy, educational attainment, labour force participation, contraceptive use, and political participation. Yet major gaps continue to exist in wages, subordination in the market, health care, nutritional support, and education.[2]

Studies have shown that women face disproportionate hardships in labour markets, wage levels, lack of access to capital, advisory services, and access to technology. Additionally, the weakening, and in some cases collapse, of traditional organizational, cultural, and security structures as a form of social capital has left them with less support for the care work expected of them. Women's vulnerability, with the additional risks they face in conflict and war zones, and as victims in the illicit drug trade and in human trafficking, requires that careful attention be given to the impact that the financial crisis is having on their lives. Importantly, a 2007 study presents transnational empirical evidence showing that a decline in the gender wage gap in the South relative to the North has a positive impact on the net barter terms of trade of the South (Van Staveren/Elson/Grown/Cagatay 2007). Thus, governments of developing countries should take an interest in reducing wage labour discrimination against women in their manufacturing sector. The study also points out that equivalent research has not been carried out in the services sector which, in fact, has had such a high profile in recent years in international trade negotiations.

Studies have also shown that the 'comparative advantages' that give countries an edge in the international market have in many sectors been based, in fact, on 'women's disadvantages', as they are vulnerable in labour conditions and have restricted access to credit, technology and markets (Arizpe/Aranda 1981; Van Staveren/Elson/Grown/Cagatay 2007).

A very important indicator is that the OECD has reported that women's labour force participation has stagnated, both for women aged 20–24 and for those aged 25–49, with wages in manufacturing lagging far behind those of men (Buvinić/Morrison/Ofosu-Amaah/Sjöblom 2008: 1). The *United Nations Development Fund for Women* (UNIFEM 2008/2009: 17–18) also reports that eight out of ten women workers are in vulnerable employment in sub-Saharan Africa and South Asia (see also Sandler 2007). These reports also indicate that one in five members of parliament are women; that one in four women die as a result of pregnancy and childbirth who could have been saved by effective access to contraception; and that there is a feminization of AIDS in Africa.

A high-priority issue is to assess progress in the Millennium Development Goals related to women. In fact, the Millennium Development Goals overlooked the role of women as producers so there has been no advance in policies aimed at avoiding discrimination and bias in women's access to productive inputs. This explains why poverty has not diminished in some parts of the world.

There is, however, a brighter side with the Millennium Development Goals' target of parity in primary and secondary education, with the OECD reporting that of 106 countries, eighty-three achieved the target in 2005, although nineteen

---

[2] A few years earlier, Marie-France Lebrecque (personal communication), in her study of rural development in Mexico, had posited a double masculine domination, as gendered institutions were added to traditional cultural domination.

countries will not meet this target even by 2015 (Sandler 2007). As is well known, better education not only leads to higher individual incomes, decline in child mortality and other benefits, it is also a necessary (although not sufficient) pre-condition for long-term economic growth (Lutz 2008/2009). Despite this progress, statistics show that 57 % of those left out of school are girls.

By the beginning of the first decade of this century, 'unfair globalization' (World Commission on the Social Dimension of Globalization 2004), had led to the 'crisis of development', a term echoing many field studies of the impact of neoliberal poli-cies on gender showing that the rich had grown much richer in a context of economic instability and growing social conflicts fuelled by inequality (Benería 2003: 6). In this context it is essential to consider the social construction of markets affected by government policies aimed at increasing capital mobility and expanding market-oriented production. Such policies do not consider how market-oriented production might impact livelihoods that sustain households or how people without any access to capital might be able to participate in production (Benería 2007). Additionally, in many cases, as shown by Agarwal (1989), women take on additional unpaid duties in assisting their husbands in projects for development (see also Pedrero 1996).

In sum, despite attempts at the end of the seventies to 'integrate women' into development, or to mainstream gender in development policies in the eighties, women's advancement shows a picture of unfulfilled promises, a situation well described in the title of (Tinker's 1990) book *Persistent Inequalities*. Embedded dis-crimination which had already created unjust distributions of power and resources between men and women has in many cases been strengthened by development policies which do not take into account their impact on women's employment, unpaid work and added care work (Sen/Grown 1988; Jackson/Pearson 1998).

The unevenness in women's advancement in global development is reflected in women's participation in cross-border migrations. Geographical mobility has opened doors for women, especially for rural women, but the outcomes of their experiences of migration show mixed results. In this chapter, reference will be made to this process, primarily in Latin America and using empirical evidence derived from fieldwork conducted by the author in rural villages in Mexico.

5.4 Cross-Border Migrations and Illicit Activities

Cross-border migrations and illicit trafficking of people is high on the list of con-cerns on the development agenda. International migrants—191 million in 2005, probably closer to 200 million if illegal migrants are included—send more than 250 billion dollars annually to their home countries (DESA 2006).[3] The economic

---

[3] Note that figures on international migration are based primarily on the number of foreign per-sons registered in national censuses and other official sources. They therefore do not include undocumented migrants, nor those involved in the trafficking of human beings, in the drug trade, or in other criminal activities. See the website at: http://esa.un.org/migration.

crisis is already bringing about a steep decline in such remittances, which have become vital for poor households in many developing countries.

The United Nations (DESA 2006) reports that the migration stock (the permanent nucleus of international migration transfers) rose from 0.8 % of the world's population in 1960–1965 to a peak of 6.7 % in 1985–1989, then fell back to 1.5 % in 2005–2007. Cross-border migrant flows have now become a global phenomenon. Data shows that twenty-eight countries have received more than one million immigrants each, and sixty-three countries have received more than 100,000. Interestingly, international migrations, overall, represent the same percentage of the world's population, 3 %, as they did at the start of the twentieth century, but with a very different composition (Centre Tricontinentale 2004).

Recent studies have referred to the 'feminization' of international migrations, since women's participation in these flows increased to 49.6 % in 2005 (UN 2006). Until the eighties, trends in international migration showed a higher gender ratio of male to female workers. The exceptions were labour-specific, for example, male labour migration to mining regions in Africa and labour migration to Islamic countries in the Middle East, or culture-specific, such as the migration of women from the state of Kerala in India, where, in contrast to Hindu states, women migrated to work as nurses and other service economic activities.

The ratio of male workers to female workers began to shift as cross-border migrations increased in the last three decades. A gender perspective is necessary to understand these shifts by ascertaining, in each case, whether a greater number of women are migrating or whether the ratio is due to fewer men becoming involved in such flows. The increase in women migrants was initially related to more women joining their spouses or families who by then were permanently established in countries of destination. In recent years, however, women began to migrate as independent workers, following a trend which was also evident in most countries' internal migratory flows.

One of the reasons for this 'feminization' in cross-border flows is the expanded demand for care work, both skilled such as nurses and unskilled such as domestic workers, in industrialized countries. This highlights the centrality of care work as the basis on which social policies can either promote gender equity or exacerbate gender divisions and inequality (Zimmerman/Litt/Bose 2006: 376).

Among skilled migrants, internationally, women lead the 'brain drain', now called the 'brain gain' in industrialized countries. Evidence for this is that the difference between the number of men and women migrants with tertiary education is 3.2 % in Latin America and the Caribbean, 8 % in Oceania, and 10 % in Africa (UNIFEM 2008/2009: 11).

The market is becoming global but the rules that migrants encounter in their cross-border movements are always national. A distinction must be made between the 'transnational' character of their travel and the national definition both of their working wages and conditions, and of their access to housing, schooling and other social services. Additionally, the situation of women migrants varies in segmented markets. For example, for women migrants who are care workers in institutions, the prevailing national laws and attitudes define their situation, but for those who are care workers in private family homes, rules are much less defined and subject to government supervision (Zimmerman/Litt/Bose 2006).

5.5 The Changing Lives of Women Migrants in Mexico

In the case of migrants from Latin America and the Caribbean, the shift from women migrating to the United States with their kin (husbands, brothers, aunts, and other close kin) to travelling independently is very marked. During the eighties, and especially the nineties, the routes for this massive flow of migrants to the '*Norte*' were set up, as family or kin members began to reside in the US. This opened up opportunities for women to live with trusted families or village friends in the receiving communities, and to find jobs. Over time, this has given women, especially younger women, confidence in making the trip on their own and striking off on their own in American society.

Since the nineties, the 'pull' factor has become more important as a motivation for young women to migrate out of their rural villages. Through education and the mass media, young girls and women have acquired new values of self-confidence, and a desire for personal achievement. To the question "Why do you migrate?", most young women answer along the lines of "*para superarme*", "to better myself". For women, however, in addition to overcoming traditional norms and expectations, the returns from education may be relatively higher in the US, because, for example, of gender discrimination in Mexican labour markets (Benería/Roldán 1987; Kanaiaupuni 2000). Although young women are seeking a better status through migration, studies have found that even though women's salaried work does tend to improve their gender relationships in the household, migration does not change the traditional gender roles (Briones 2002: 183).

In the last decade, however, the 'push' factors in female migration in Latin America and the Caribbean have also become stronger in many countries. Neoliberal policies that have undermined traditional agricultural livelihoods and economic growth with high levels of inequality and unemployment have worsened the situation of women in low-income families in many Latin American and Caribbean countries. Another factor influencing women's outmigration rates is the increased monetary needs of their families as government social services have diminished.

In Mexico, liberalization policies have practically destroyed smallholding agriculture, forcing millions of rural people to migrate from rural areas. This has worsened the situation of women, many of whom have to migrate either to cities, to the United States, or on to the seasonal circuits of agricultural labourers. Working as agricultural labourers is perhaps the most debased and exploitative work situation for everyone, but especially for women and children, as my own fieldwork research has shown (Arizpe/Aranda 1981; see Chapter 7 "Women Workers in the Strawberry Agribusiness in Mexico" in this book). Many studies confirm this degradation in which, as Sara Lara explains, agribusiness employers are able to exercise domination over women migrants, many of them indigenous women, "… through threats, deception, pressures and harassment but most of all by humiliating them. This finally has the same effect as direct violence. But, additionally, this unleashes the worst sentiments towards the 'other' among the workers themselves….allowing, at the same time, the most brutal expression of violence inside the families" (Flores 2003: 394).

For women it takes courage to migrate now that the darker side of migration has grown more dangerous in many regions of Mexico, especially the border areas, in recent times. Indeed, on the Mexican border the toll of 'feminicides'—a new type of homicide, now defined in Mexican law as a hate crime against women, typically perpetrated by the husband or a close kinsman—has reached more than 600 women killed. This number has greatly increased in the last decade because of the federal and state governments' total incapacity to investigate, arrest, and punish the perpetrators. This impunity has led to young women becoming the victims of a horrendous web of human traffickers, drug '*sicarios*' (assassins), perverts (as in 'snuff' films in which women are actually killed), and corrupt police and government officials.

It is important to mention that when the doors mentioned above are closed to young women (jobs, migration to cities, migration to the US), they then have two other options. One which has already been mentioned is political militancy as guerrillas, as in the case of the Zapatista army in Chiapas.

The other is the grey and black economy. This may mean work in the grey 'entertainment' industry, such as in 'table dance' bars which have proliferated along the border, or down the slippery slope, voluntarily or involuntarily, into prostitution and illicit human trafficking. Or, increasingly, since this rarely happened a decade ago, into direct involvement with the drugs trade. It must be said, though, that in many cases women are inducted into such activities after falling in love or being seduced by men displaying wealth and power and promising these young girls things they had never dreamed of. In exceptional cases, such love affairs have ended with women becoming heads of drug circuits after their husbands or lovers are killed. One such woman is Sandra Ávila Beltrán, '*la Reina del Pacífico*' (the 'Queen of the Pacific'), who at one point controlled the drug supply routes over the Pacific, was jailed in Mexico without being extradited to the United States and is now free.

As to the women who stay in the rural household while their husband or children migrate, it is worth noting that their precarious situation will vary according to the income level of the family, access to inputs for productive activities, and the support of broader kinship networks. The presence of the female household head is crucial for maintaining the integrity of the household for the migrant men: "For the typical migrant labourer who prefers not to remain permanently in '*el Norte*' it is much easier to negotiate a strange country with a wife at home sustaining the family and social relationships. Having a wife at home is cost-efficient, conforms to the gender norms, and also enables him to move back and forth without losing social standing in village and kinship structures" (Kanaiaupuni 1995, 1998, 2000: 1319–1329). This does not preclude his having a second family, often with another Latino woman, with whom, in the end, he stays. Whether or not he continues to send remittances back to his first family, or to keep up his ties with his children by his first wife, will depend on his affective and psychological make-up, but will be greatly influenced by his cultural background. This is why I have always argued that remittances are a cultural phenomenon.

As to Mexican women migrants' experiences in the United States, (Hagan 1998, cited in Parrado/Flippen 2005: 608–609) gives the broad setting of their employment: "The sex segregation of migrant occupations also limits the networks of migrant women. Hispanic migrant men tend to concentrate in construction, manual labour, and services in which they have extensive and varied contacts with other men. Hispanic migrant women, on the other hand, are concentrated in domestic and small-scale service occupations, which are more isolated and have fewer avenues for advancement than the occupations of migrant men".

Several studies suggest, however, that women-based networks can transform the female migrant experience, providing links to employment, assistance, and information in destinations (Donato/Kanaiaupuni 2000; Hondagneu-Sotelo 1994). In their 2005 study on migration and gender among Mexican women, Parrado and Flippen challenge an assimilationist, emancipating view of migration and gender that would predict a gradual and unidirectional increase in Mexican women's power associated with migration and US residence. Instead, they argue that "...the effect of migration on gender relations is highly variable, with gains in some realms offset by losses in others. In keeping with the selective assimilation literature, we find that Mexican migrants selectively incorporate some aspects of the receiving society while simultaneously reinforcing cultural traits and patterns of behaviour brought with them from their communities of origin. Although this process of adaptation without assimilation may help insulate migrants from the destabilizing forces arising from residence in a foreign environment, the end result is that in some instances, migration actually exacerbates gender imbalances" (Parrado/Flippen 2005: 626–627).

More generally, they also conclude (2005: 606–632) that the association between migration and gender relation is not uniform across different gender dimensions: "the reconstruction of gender relations within the family at the place of destination is a dynamic process in which some elements brought from communities of origin are discarded, others are modified, and still others are reinforced. Results challenge the expectation that migrant women easily incorporate the behaviour patterns and cultural values of the United States and illustrate the importance of selective assimilation for understanding the diversity of changes in gender relations that accompany migration".

A further important conclusion (2005: 628) is that "Prior theorizing about the effect of migration on gender has in many cases portrayed the vestiges of traditional gender arrangements from communities of origin as an important constraint on migrant women's socioeconomic advancement. Contrary to this interpretation, our research suggests that the causal connection is likely to work in the other direction as well. It is not that migrant women fail to 'progress' toward more egalitarian norms because of their cultural background or patterns of behaviour brought from their communities of origin. Rather, it is their structural position within the US society including their precarious legal status, unfavourable work conditions, and lack of social support that undermines their well-being and power within relationships".

In spite of all these hardships, as Sara Lara concludes, migration creates the possibility of new scenarios in which gender and cultural structures may be altered and in which women, especially indigenous women "are able to generate spaces of interaction that give them weapons for a new counter power that they do not have in their villages of origin" (Flores 2003: 395).

5.6   Monitoring the Impact of the Economic Crisis on Gender and Cross-Border Migrations

To monitor the impact of the present crisis on women in the field of international migrations, it is important to continue to develop research on the following topics:

1. *The decline in remittances from migrants*

   Declining remittances will have serious consequences in households that use remittances for day-to-day living expenses, especially if women have been left as heads of households and may have to expand their income-generating activities in the informal sector. If remittances decrease, then investment in small business or trade activities in their regions of origin will slow down.

2. *Return migration*

   The number of returning migrants to Mexico, as a result of the decline in jobs in the construction industry and, generally, of the economic recession in the US as well as the patrolling of border areas and the hardening of migratory regulations in the US, will probably reach several million. Many of these migrants may use the skills they learned in migration to set up small businesses, shops, or trading activities, but they would need credit support and either government extension services or private support for small businesses to be successful in such initiatives. Mexican migrant women who were employed in the domestic care economy will not have acquired new skills, but hopefully, they will have been able to set aside savings for their return home. Migrant young women, whether in the domestic or industrial care economy or in other jobs, may have acquired new skills, but most of them will also have increased their aspirations as to jobs, incomes, respect from men, and protection from domestic violence. This is especially and poignantly true of those born to families of Mexican illegal migrants who have gone through elementary and high school studies and who have already absorbed the achievement habits of American youth. It is for this reason that programmes for amnesty or regularization of those in such a situation is a priority.

If urgent measures are not taken to create employment and educational attainment for returning Mexican migrant youth, especially in the deprived rural regions from which they first tried to escape, there is the risk that illicit activities—drugs, organized and petty crime, and prostitution—may again expand. According to research carried out by my students, those in danger of being deported may cling to their clandestine situation and turn to the black economy, always connected to cross-border networks.

If young returning or deported migrants land back into hopeless situations, where neither jobs nor educational opportunities nor any other window is open to them, apart from the black economy, they will hide their anxieties through mindless violence, taking drugs or, in case of young women, becoming pregnant and forever living on the margins of poverty, violence and suicide.

Those women who still try to keep their families and kin and life together will bear the brunt of these criminal and tragic events as family caretakers and as victims of abductions and domestic violence. Also, as outright rage or, more usually, as the level of internalized rage that becomes depression rises, they may break with the passivity of the traditional female construct and become willing and active participants in the drug and prostitution trade, or in abductions.

Social policies that comprise only handouts to just a few selected mothers in towns and villages or to just one of their offspring whose selection is never fully explained—where other mothers know and seethe about it!—are totally useless in stopping the fraying of the social fabric and the destruction of young lives. Had the money that the Mexican government spends fighting the drug trade and organized crime been invested in quality education, in supporting small enterprises and in protecting women from violence, the brutal decay of many Mexican regions and the ruined lives of its 'demographic bonus'[4] would have been avoided.

5.7 Conclusions

This review of the literature on women's advancement in international development in recent years shows areas of improvement of women's lives and also regression due to the stagnation of their status or to an increase in their work burdens or in risks of violence and criminality. Quite significantly, the same trend is evident in the review of recent research on women's international migrations and in my own research observations. Parrado/Flippen (2005: 609) put it succinctly when they say that "…Taken together, the broad literature on migration and gender highlights the need to distinguish between domains in which migration has led to gains in women's autonomy and those in which inequities are maintained or reinforced". To allow for such an analysis it is very important, then, to treat gender as a "…theoretical basis of differentiation and not simply as a control variable…" (Kanaiaupuni 2000: 1312).

Given the inequality and exclusion that has been fostered in the last two decades through economic growth, it is vital to define development as the growth of markets which are embedded in the broader political and social ideals of societies. Women's advancement will depend on whether they are included as a constitutive part of these ideals and their outcomes. Consumers must be simultaneously theorized as citizens of a nation state and as members of social communities. The State must be strengthened in order to curb abuses by financial and business enterprises; this includes paying attention to obesity and health issues, especially in children,[5] and to levelling the fields of opportunity in education, employment and investment for small income-generating activities. Very importantly, the State must provide opportunities for unemployed youth, and must have the necessary autonomy to stop the infiltration of organized crime into all sectors of society.

---

[4] The 'demographic bonus', in Spanish '*bono demográfico*', is the bulge in demographic growth in many developing nations occurring just after the demographic transition and which, in most countries, is seen as a dynamic way to have young people promote economic growth.

[5] In the last decade, Mexico has risen to one of the highest levels of child obesity in the world.

Given the economic recessions, have women been able to influence the international development agenda? The authors of the book *Developing power: How Women Transformed International Development* (Fraser/Tinker 2004) argue that the Women in Development movement has taught women in and out of government to think and act differently in virtually all fields of activity.

Again, however, studies have shown that analysing the impact of development on women requires new theoretical tools to address contradictory and uneven outcomes. Nancy Fraser defines this challenge as a dilemma posed by the 'post-socialist' condition in capitalism. She defines the 'post-socialist' condition as "…an absence of any credible overarching emancipatory project despite the proliferation of fronts of struggle; a general decoupling of the cultural politics of recognition from the social politics of redistribution; and a decentering of claims for equality in the face of aggressive marketization and sharply rising material inequality" (2004: 1103–1124). She proposes a model of radical democracy that will "…combine the struggle for an antiessentialist multiculturalism with the struggle for social equality" (Fraser 2004: 1105).

As a final comment, I believe Nancy Fraser has expressed the issue very aptly: "In seizing this moment, we might just bend the arc of the impending transformation in the direction of justice—and not only with respect to gender" (2009: 117).

References

Agarwal, Bina, 1989: *A Field of One's Own: Gender and Land Rights in South Asia* (Cambridge: Cambridge University Press).

Arizpe, Lourdes; Aranda, Josefina, 1981: "The Comparative Advantages of Women's Disadvantages: Women Workers in the Strawberry Agroindustry in Agribusiness in Mexico", in: *Signs*, 7,2 (Winter): 453–473.

Benería, Lourdes; Roldán, Marta I., 1987: *The Crossroads of Class and Gender: Homework, Subcontracting, and Household Dynamics in Mexico City* (Chicago: University of Chicago Press).

Benería, Lourdes, 2003: *"Gender, Development and Globalization: Economics as if People Mattered"* (New York: Routledge).

Benería, Lourdes, 2007: Gender and the Social Construction of Markets, in: Van Staveren, Irene; Elson, Diane; Grown, Caren; Cagatay, Nilufer (Eds.): *The Feminist Economics of Trade* (London: Rouledge): 13–32.

Budlender, Debbie, 2008: *The Statistical Evidence on Care and Non-Care Work across Six Countries* (Geneva: UNRISD).

Buvinić, Mayra; Morrison, Andrew R.; Ofosu-Amaah, A. Waafas; Sjöblom, Mirja (Eds.), 2008: *Equality for Women: Where Do We Stand on Millennium Development Goal 3?* (Washington DC: IBRD/WB).

Centre Tricontinentale, 2004: *Genèse et enjeux des migrations internacionales* (Paris: Syllepse).

Chun, Dorothy E.; Boyd, Susan B.; Lessard, Hester, 2007: *Reaction and Resistance: Feminism, Law and Social Change* (Vancouver: UBC Press).

Deen, Thalif, "Development–UN: Recession threatens Women's Meagre Gains", in: *Global Information Network*, 4 March 2009: 3.

Donato, Katharine M.; Kanaiaupuni, Shawn Malia, 2000: "Poverty, Demographic Change, and Women's Migration from Mexico", in: García, Brígida (Ed.): *Women, Poverty and Demographic Change* (New York: Oxford University Press): 217–242.

Fraser, Arvonne S.; Tinker, Irene (Eds.), 2004: *Developing Power: How Women Transformed International Development* (New York: CUNY Feminist Press).

Fraser, Nancy, 2009: "Feminism, Capitalism and the Cunning of History", in: *New Left Review*, 56 (March–April): 97–117.

Fraser, Nancy, 2004: "To Interpret the World and to Change It: An interview with Nancy Fraser", in: *Signs*, 29, 4 (Summer): 1103–1124.

Hagan, Jacqueline Maria, 1998: "Social Networks, Gender and Immigrant Settlement: Resource and Constraint", in: *American Sociological Review*, 63, 1: 55–67.

Harcourt, Wendy. 2009. "Editorial: Women's Global Organizing", in: *Development*, 52: 133–135.

Harris, Anita, 2008: *New Wave Cultures: Feminism, Subcultures, Activism* (London: Routledge).

Hondagneu-Sotelo, Pierrete, 1994: *Gender Transitions. Mexican Experiences of Immigration* (Berkley: University of California Press).

Jackson, Cecile; Pearson, Ruth, 1998: *Feminist Visions of Development: Gender Analysis and Policy* (London: Routledge).

Kanaiaupuni, Shawn Malia, 1995: "Gendered Migration Decisions: The Decision to Migrate among Mexican Men and Women", Paper presented to the American Sociological Association (ASA), August.

Kanaiaupuni, Shawn Malia, 1998: "Invisible Hands: Women's Participation in U.S.-Mexico Migration Patterns", Paper presented to the American Sociological Association (ASA), San Francisco, California.

Kanaiaupuni, Shawn Malia, 2000: Reframing the Migration Question: Men, Women, and Gender in Mexico, in: *Social Forces*, 78, 4: 1311–1348.

Lara Flores, Sara María, 2003: "Violencia y contrapoder: una ventana al mundo de las mujeres indígenas migrantes en México", in: *Revista de Estudos Feministas*, 11, 2 (July–December): 381–397.

Lutz, Wolfgang, 2008/2009: "New Data on Human Capital", in: *Population Network Newsletter*, 40 (Winter): 1–8.

Moghadam, Valentine M (Ed.), 1996: *Patriarchy and Economic Development: Women's Positions at the End of the 20th Century* (Oxford: Clarendon Press).

Parrado, Emilio A.; Flippen, Chenoa, 2005: "Migration and Gender among Mexican Women", in: *American Sociological Review*, 70, 4 (August): 606–632.

Pedrero, Mercedes, 1996: *Mujeres y Empleo en el Contexto del Desarrollo* (Mexico City: CRIM-UNAM).

Sandler, Joanne, 2007: "Gender Equality is Key to Achieving the MDGs: Women and Girls are Central to Development", in: *UN Chronicle*, 44, 4 (December): 47–8.

Scott, Catherine V., 1995: *Gender and Development: Rethinking Modernization and Dependency Theory* (Boulder, Colorado: Lynne Rienner Publishers).

Sen, Gita; Grown, Caren, 1988: *Development, Crises and Alternative Visions: Third World Women's Perspectives* (New Dehli: DAWN).

Tinker, Irene. 1990: *Persistent Inequalities: Women and World Development* (Oxford: Oxford University Press).

UNIFEM (United Nations Development Fund for Women) 2008/2009: *Progress of the World's Women 2008/2009: Who Answers to Women? Gender and Accountability?* (New York: UNIFEM).

UN, DESA (Department of Economic and Social Affairs), 2006: *Trends in Total Migrant Stock: The 2005 Revision* (New York: UN).

Van Staveren, Irene; Elson, Diane; Grown, Caren; Cagatay, Nilufer (Eds.), 2007: *The Feminist Economics of Trade* (London: Rouledge).

Vega Briones, Germán, 2002: "La migración mexicana a Estados Unidos desde una perspectiva de género", in: *Migraciones Internacionales*, 1, 2 (January–June): 179–192.

World Commission on the Social Dimension of Development, 2004: *A Fair Globalization: Creating Opportunities for All* (Geneva: International Labour Organization).

Zimmerman, Mary; Litt, Jacquelyn; Bose, Christine, 2006: *Global Dimensions of Gender and Carework*. (California: Stanford University Press).

Chapter 6
Feminism: From the Outcry of the Seventies to the Strategies for the Twenty-First Century

6.1 Introduction

When one lives a historical experience, it is only much later one can understand the broader significance of what happened. This has happened with feminism in Mexico, especially for those of us who acquired "a conscience of one's own" when feminism spread in the 1970s.[1] As we look back, we now have different interpretations of what we experienced because feminism has evolved and, with the wings of history, we ourselves have also changed. Who are we today: post-feminists or feminists with a new perspective? I believe we are feminists with a new perspective, because living in a democracy makes us shift strategies while keeping intact the original dream of having a room of our own, a conscience of our own, and a life that is worth living.

We must not stop now. We must look toward the past, but with tenderness, in order to understand the fears and misapprehensions that made the road we travelled and the bold actions we had the courage to embark upon difficult. Let us begin by stating that we feminists tried to live according to our aspirations in whatever way we could. By doing so, we won a lot; but the road is long and we still have a distant vanishing point on the horizon to reach and another one farther still to be reached by the young women who today look out towards it with a luminous gaze.

6.2 On the Margins of Being

As always happens when we look to the past, it seems inconceivable that women had not always lived with a consciousness of self and freedom of choice, but this is the way things were before the sixties in Mexico. It could not be otherwise in

---

[1] This text was originally published in Spanish: Arizpe, Lourdes, 2002: "El Feminismo: del grito de los setenta a las estrategias del siglo XXI", in: Gutiérrez Castañeda, Griselda (Ed.): *El Feminismo en México. Revisión histórico-crítica del siglo que termina* (Mexico City: UNAM—Programa Universitario de Estudios de Género): 63–70.

L. Arizpe, *Migration, Women and Social Development*,
SpringerBriefs on Pioneers in Science and Practice 11,
DOI: 10.1007/978-3-319-06572-4_6, © The Author(s) 2014

Poster of the First Mexico-Central American Symposium on Women Studies, Mexico City, of which Lourdes Arizpe was Secretary-General, Mexico City, 1977. This poster is from the author's personal archive and was photographed by the author

a country which was trapped in an external Cold War, but also in a silent internal one. There was much talk of revolution, but what happened was repression; women were shut into private spaces, subject to religious and sexist obscurantisms. Social peace was maintained through a repression that allowed neither freedom of speech nor freedom of association. Although women had played an important role during the first periods of the post-revolutionary regime, they were later marginalized in public life. *Machismo* was a part of the image and practice of an authoritarian regime that instead of making us responsible citizens turned us into beings with no judgement or right to think. Women could have no other vocation than to bear children, and, therefore, the old educational scheme for women was built around controlling their sexuality and fertility.

That monolithic society fractured in 1968 with the student movement, when for the first time a demographic surge of young people rejected the political and social system. But we were brutally silenced on the second of October, 1968. It was a betrayal. Many young people who were to build the nation's future were assassinated. What was left after the blunt political incompetence of the government was the indignity of silence but also our lifelong commitment to changing the regime towards democracy.

And we, young women, had now awakened.

In my own case, I had already been awakened by the National School of Anthropology and History (ENAH) and by my fieldwork in a Nahua village in the Sierra de Puebla where I had encountered the grandeur of Indian culture linked to the depths of deprivation. With this heavy political and intellectual legacy on my shoulders, I journeyed to London in 1970 to pursue postgraduate work in anthropology. I came back in 1973, with a broad vision that was fully engaged with feminism, with the world, and with a new radicalism that I believe was emblematic of the first 'global' generation. That is, the demographically more numerous generation, better educated and more informed about what was going on in history, with a naive proclivity towards thinking that the best ideals for humanity could be achieved simply by having clear ideas and a generous heart.

Another very important factor in the emergence of the Mexican feminist movement was that, after 1970, the government reacted to the tragedy of the assassination of over 300 students in Tlateloco with a 'democratic opening' exclusively... *for men*! We women students, those of us who had shared the ideas, sit-ins, demonstrations, repressions and persecutions, continued to be invisible. It also became apparent, as visible as blood splashes in the public square, that although women were unequal in everything else, they were equal when it came to disappearances, torture and murder, all in the name of social peace. It is not surprising, then, that we began to create our own spaces for public action.

Thus, the 'little groups', as we called them, were born, some of them tending more towards political action and others toward more consciousness-raising initiatives. A very long list could be given of the activities of many of these feminist groups in Mexico in the seventies. In my case, I was dazzled by the lecture given by Susan Sontag in the Faculty of Political and Social Sciences at the National University, followed by the very liberating talk we had with her after her lecture, when we sat on the benches and lawns of the university campus. It was there that we began to weave together the threads that would later lead us to form a feminist group in Magnolias Street.

6.3 In Search of "a Conscience of One's Own"

This was the title of my inaugural speech at the opening of the First Symposium on Women's Studies held in Mexico City in 1977. In my view, it reflected the very intense need we had to create a new vision of ourselves and of society and which led us to organize the beginning of the feminist movement in Mexico, inventing as we went along. Many women students had participated in consciousness-raising groups in the US and in Britain, where feminism had again caught fire. It was being referred to then as the 'second wave' of the women's movement, nourished by the history of the suffragists in the UK in the late nineteenth century and by the new books on 'sexual politics' and 'the feminine mystique' published in the United States.

In Mexico, the events of 1968 had pushed us to an extreme radicalism, with a political commitment especially to those who were marginalized. But we were caught in the tormented ambiguities of the 'sexual revolution', which, as has been said many times, happened more in the rhetoric than in practice. What is certain is that a critical mass of middle class and university young women had by then become involved with feminist movements, many of them linked to other parts of the world and who were intent on firmly participating in Mexican public life.

6.4 The First Global Movement: Feminism

Was it a coincidence that all of these feminist movements arose simultaneously in so many countries? Is this perhaps explained in part by the demographic and social bubble of the Woodstock generation? Why did women everywhere begin to mobilize? It is very important to know the answer to this question, since it points to the irreversibility of the social processes of recent decades. From an anthropological perspective, one of the fundamental trends in all human cultures had always been the protection of women's fertility. Of course this had to be so, since the survival of human groups in famines, diseases and genocidal wars is achieved by the capacity to reproduce. The surprising thing is that human civilizations have been able to avert these dangers. Now, on the contrary, it is surprising that curbing population growth has become an imperative for this survival.

A new window of opportunity has opened up in societies, at least in westernized societies, and this is that *by having fewer children, a large part of the lives*

*of women has been liberated so that their abilities can now be applied to benefit
other aspects of those societies.*

Since this trend is irreversible, it has brought about maladjustment in gender
relations, in the structure of employment and income, in population settlement
patterns, and in migrations. However, in the long term, a balance will be
achieved again through self-organization. In this context, feminist movements,
along with others, have become precisely the forces that are searching for these
new adjustments. *It is therefore very important to recognize that this search
must continue to make constant adjustments in its own strategies.* Thus, as
occurs in any other social movement, the feminist movement needs to adjust its
strategies in accordance with the shifts in the economic and political context in
which it evolves.

The feminist movement cannot be conceived, therefore, just as a change of
'values' or 'lifestyles', because it is strongly enmeshed with the dynamics of the
population, the economy, politics, and culture. Furthermore, we now also know
that these evolve within the framework of the carrying capacity of the biosphere
and the geo-atmosphere. *Our challenge is to understand all these relations again
today.* Culture and gender relations, as could be expected, are part of new processes that are global, since they affect all the inhabitants of the planet.

Understood in this way, what happened to us as the first conscious feminists since the seventies takes on many more implications. On the one hand, the
official and non-official history of Mexico still has large black holes regarding
the history of women. With the exception of Sor Juana Inés de la Cruz and the
Corregidora Josefa Domínguez, women seem to have never existed. As women
suddenly appeared from the most surprising places, wanting to write in our feminist journal, *Fem,* we happily discovered that there was an abundance of cou-
rageous, frustrated, hopeful and enraged women in our present who wanted to
change the past. With *Fem* we were able to open up a crack that would lead us to
freedom of choice.

6.5   "The Road is Built by Walking"

In the beginning, my feminist group was oriented towards 'consciousness-raising'.
What drove us was a deep feeling of fraternity in political actions with other students and of *sisterhood* in feminist groups. I shared this with friends who had
refused to fight in Vietnam, friends who had gone to meditate in an *ashram* in
India, and with freedom fighters in many decolonizing countries.

I found myself propelled into anthropological fieldwork, driven by anger derived
from my own experiences and at the abuse suffered by poor women in Mexico. I
was enthusiastic about opening up new avenues of thought that would open doors
for peasant and indigenous women. In 1975, I published a book on anthropology about the 'Marías', the Mazahua indigenous women who had appeared in the
streets as if to tear away the thin veil of development centred on Mexico City.

By then I had just finished my doctoral dissertation in anthropology for the London School of Economics about the migrations of peasant families to Mexico City. My fieldwork and all the interviews showed that this uprooting was going to upset the whole social structure and cultural ways of life, both in the communities and in the cities. I also knew, from a vast sociological literature, that urbanization and industrialization would change the composition and functioning of families and therefore the ways of life of women and men. This would happen regardless of any social movement such as feminism. To be engaged in research action at that time, I told myself, could help to stop the violence, rape and abandonment of women which the women wanted to rebel against though they were unable to do so. We had been naïve but we threw ourselves into trying to improve things, and today, two decades later, women's rights and opportunities have unquestionably improved in Mexico, although never to the extent we would have liked them to.

It seems to me that much of the influence we were able to achieve was through the hopeful, difficult launching of *Fem*. This feminist magazine opened up a channel that touched the lives of most Mexican women. Alaíde Foppa had invited me to a meeting in her home in which the project was born. Although we lived this experience through unbelievable complications, the journal made strides, as decisive as they were discontinuous, with pieces and remnants that seemed at times to be woven together only with tenacity. And with pure friendship, that special thing that grows among pioneers who live dangerously and survivors who will never forget those who silently disappeared. This is the tribute we all continue to give to Alaíde Foppa, who was kidnapped and murdered by the Guatemalan secret police.

6.6 Walking On

I will now jump ahead to the end of the twentieth century. In the nineties I had to face the dilemma of sustaining, at the same time, an active feminist militancy and strenuous professional work which had to be doubly efficient to be accepted in the same way that the work of male colleagues was accepted. I began to hold high elective and organizational positions, many in international organizations, in which it became evident to me that to do the work entrusted to me without falling into a 'masculine style of exercising power', I would have to develop a cluster of subtle intellectual and political skills that we women had never been educated in.

Indeed, women of my generation had to "build the road as we walked". There were few women in previous generations that could be professional or political models; we had to hold on to our positions and exercise power better than our male colleagues, learning as we went along the way; and what is more important, we have had to *learn to understand* our womanly ways of analyzing, managing and taking decisions—by trial and error, since part of this way definitely did not have successful outcomes—and then applying them consciously even if our fellow workers were surprised.

I would say, therefore, that we pursued a 'triple militancy'. There is the feminist 'dual militancy': that of a political participation with the militancy of a new type of presence, as women at the centre of processes of change in our society. Then, a 'third militancy' is added as women have become decision-makers in government, corporations, political parties, media, trade unions, and civil society. In all of them, a critical mass of women have now taken leadership and are pushing hard to break through the glass walls and ceilings, creating the new fabric of democracy in Mexico.

Conference on Gender and Equity, Society for International Development Conference, New Delhi, March, 1988. This photo is from the author's photo collection

Chapter 7
Women Workers in the Strawberry Agribusiness in Mexico

Lourdes Arizpe and Josefina Aranda

7.1 Introduction

In recent years, the women's movement the world over has stressed the need in recent years, the women's movement the world over has stressed the need to provide women with increased access to salaried employment in order to improve their living conditions. In some industrialized countries, however, the recession and long-term economic trends are making it more difficult for women to get adequate employment, because, among other reasons, many of the jobs traditionally held by women in industries—particularly in textiles, garment manufacturing, and electronics—are being relocated in developing countries (UNIDO 1980). In some cases, many of the labour-intensive agricultural activities in which women worked as wage labourers have also been shifting to developing regions. In the latter, as the economic structure maintains high levels of male and female unemployment, most governments welcome capital investments that will create employment and bring in foreign currency through exports.

Behind this movement lie both the market pressures that forced companies into a constant search for lower production costs, and the rationale of 'comparative advantages,' according to which different economies are advised to specialize in those products that they can sell profitably in the international market. But it so happens that in the case of women, such 'advantages' are closely linked to the cheap labour costs that come from the discrimination against them in the labour market. Thus, it could be said that women in developing countries are gaining the jobs that have been redeployed from industrial countries. In fact, companies are using women's

This text was originally published as: Arizpe, Lourdes; Aranda, Josefina, 1981, in: *Signs*, 7, 2, University of Chicago (Winter): 453–473. Permission was granted on 12 June 2013 by Mgr. Perry Cartwright for Chicago University Press.

The research underlying this chapter was financed by the Rural Employment Policies Branch of the International Labour Office in February 1981.

liberation slogans in deprived areas to justify giving jobs to eager young women rather than to older women or men who also desperately need jobs (Lim 1978).

The main issue raised by these events—whether the fluidity of the international labour market has become more of a Zero-sum game for women than for men—cannot be fully discussed in this chapter, but some light can be shed on it by examining the extent to which such a 'gain' for women in a developing country actually improves their status and living conditions. A survey through interviews of young Mexican peasant women who have recently entered salaried employment in the strawberry-export packing plants of Zamora in the state of Michoacán helps us to understand the changes created by salaried work in their consciousness, their living conditions, and their situation within family and community.

7.2 Agroindustry and Rural Employment in Developing Countries

Worldwide, the optimism generated in the 1950s by the projects for rural community development and, after that, by the increase in agricultural production due to the Green Revolution, came to an end in the 1970s. Meanwhile, in the last three decades rural unemployment, movement of peasants toward the cities, demographic growth, and the marginalization of rural women from the technological and economic benefits of development have increased rapidly in many countries of Latin America, Africa, and Asia.

Import-substitution policies as a strategy for development in such countries led to rising foreign debts due to the high costs of technology and of capital goods imported from the industrialized countries (Todaro 1977). The governments of developing countries, in order to acquire foreign exchange to improve their balance of payments, have encouraged export-oriented agriculture, which has led to food scarcity in many rural areas of Africa, Latin America, and Asia (Moore/Collin/Fowler 1977). Attempts to compensate for this scarcity by purchasing food from abroad have only perpetuated the vicious circle of dependency and poverty (George 1977).

Technological improvements from the Green Revolution increased yields and efficiency in rural production, but also led to higher concentration of agricultural resources in the hands of capitalist entrepreneurs (Palmer 1977; Hewitt de Alcántara 1978). In many countries this concentration has displaced small family producers who have become agricultural labourers or migrants surviving precariously in the outskirts of overpopulated cities (Arizpe 1978). The expansion of this surplus population in rural and urban areas is being attacked through massive family-planning campaigns, even though it is clear that population growth is closely linked to the conditions of extreme poverty and insecurity that prevail on the land. Another solution now being proposed to stop the rural exodus lies in the creation of rural employment through agroindustries, a policy sponsored both by national governments and by multinationals who have found a fertile field for investment.

Following this trend, in Latin America the per capita production of subsistence crops decreased by 10 % between 1964 and 1974, while that of agricultural products for export increased by 27 % (Burbach/Flynn 1978: 5). During the same period US capital investments in agriculture for export in this region increased considerably, since investments in the food industry provide a 16.7 % profit abroad, compared to an 11.5 % return within the United States (Feder 1977). Since World War II, food-processing companies have invested more in Mexico than in any other country of the Third World. An example of this type of investment is the strawberry industry in Zamora, which since 1970 has provided employment for approximately ten thousand young peasant women in its packing plants. As in the textile and electronics industries, which are also redeploying their production units abroad, the employment of women rather than men is clearly preferred in these agroindustries (Lim 1978; George 1977). Why are young women preferred? Is it sufficient to say, as do the managers of such plants, that it is because they are 'more dexterous' and 'less restless'?

7.3 Peasant Women and Rural Development in Latin America

According to recent census statistics in Latin America, women's agricultural work shows a relative decrease in all countries and an absolute decrease in many (ILO 1980). Partly, this may be due to inadequate census registration of rural women's activities, but it also reflects increased female migration from rural areas, as well as the shift to other self-employment (especially petty trade), and intermittent domestic service—occupations that fall between the holders of organized economic activities and unpaid female domestic and community work (UNDP 1980; Massiah 1980). Another important shift in rural women's activities has been reported among small family producers, where the agricultural labour of household women is intensified in order to increase or maintain productivity in deteriorating market conditions (Deere/León de Leal 1981). Finally, a fourth trend is also becoming widespread—wage labour in agricultural and livestock production or in agro industrial activities entered by poor, rural women (Díaz Ronner/Muñoz 1978: 327–334; Silva de Rojas/Corredor de Prieto 1980; Medrano 1980; Roldán 1980).

These four trends appear separately or in combination in different countries and regions. But all of them stem from the same process: the economic crisis of small peasant family production in rural areas throughout Latin America. Discussion of the causes of this crisis is beyond the scope of this chapter, but the major trends in the status and employment of poor rural women in Latin America must be understood in the context of strategies these households use to survive in an increasingly difficult environment. There are also, of course, large numbers of women who have broken completely with their parents' or their husbands' households and who live and make decisions on their own. We find them, for example, along the Mexican-US border or in the shantytowns of all the major Latin American and Caribbean cities (Fernández-Kelly 1983). Their choice of economic activity and life-style constitutes an individual decision-making process that should be analysed as such within the narrow limits set by widespread unemployment and underemployment, cramped housing, and strict social pressures.

But in agrarian societies, there is little room for individualistic response. Especially in the case of young peasant women, the decision to work or to migrate is either made by the family patriarch or through permission granted by him. In any case, even more than sons, daughters are bound to their parents' households by religious and social norms that prescribe absolute obedience, docility, and service toward others. In fact, we argue in this chapter that it is precisely these qualities that make young women so attractive a work force. The data that follow should make this abundantly clear.

7.4 Strawberry Packing and Freezing Plants in Zamora

The strawberry agribusiness in Zamora began to expand in the mid-60s, first through US capital and later through Mexican capital. Its competitiveness in the international market comes from the fact that Mexican strawberries are cultivated in the winter and that their transport and especially their labour costs are very low (Feder 1977). Production is completely dependent on US companies: the seedlings are imported from California; the export trade is handled entirely by six US commercial brokers who have stopped attempts by Mexican plants to sell directly to the European market; and the strawberry prices are dictated by conditions in the US market, especially by the success of the California strawberry harvest.

Eighteen packing and freezing plants for strawberries functioned during the 1970–1980 cycle in Zamora and in Jacona, a neighbouring village. Among them the hiring characteristics and working conditions for women, as well as male personnel, vary little: for example, some pay $14.70 (US $.66) per hour of work on the conveyor belts and others $14.00 (US $.63) but the lower wage is counterbalanced by payment of bus tickets and by better treatment for the workers. As Marta Rodriguez, put it:

> X is the packing plant where women workers are treated the worst, and that is why they have many problems in hiring people. Even though they pay more there, the girls prefer better treatment, such as they get at Bonfil, where no overtime or commissions are paid. At X the bosses are almost Nazis.

It is interesting to note that firm X is the one that consistently shows the highest productivity and efficiency; it is the only one, for example, that has devices under the roofing to prevent swallows from nesting there. In most of the plants there is a minimum investment in installations: they are prefabricated metal structures that can easily be dismantled. Everything reflects short-term investment.

At the strawberry packing plant in Zamora, 1978. This photo is from the author's photo collection

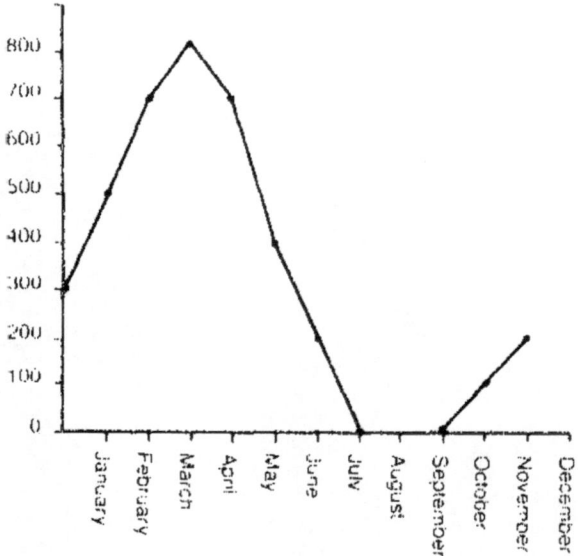

Fig. 7.1 Annual fluctuation in number of female workers employed in the Frutas Refrigeradas plant

Fifty per cent of production for export in the 1978–1979 cycle, which produced 88 million pounds of frozen strawberries (though the official figure given for exports was lower, 72.7 million pounds), was handled by the six companies we studied. Of these, three hires as many as 900 women workers at the height of the season, one hires 650, and two hire up to 350. One of the worst conditions of work women face in these plants is the acute annual fluctuation in labour demand according to changing conditions in cropping and in the price of sugar (sugar is added to the frozen strawberries). Figure 7.1 shows the typical annual fluctuation in Frutas Refrigeradas, one of the companies surveyed. Except in special conditions, all plants are closed from 4 to 6 months each year and have a peak season for hiring from March to May. Later on we shall see how the hiring is organized and how the women workers adjust their working lives to such conditions.

7.5 The Situation of Women Workers in the Strawberry Plants

Approximately 10 % of the personnel in the plants do administrative work; of these usually all managers and accountants are men, and the secretaries are young, single women from the town of Zamora. In production work, except for the young men who unload the strawberry crates from the trucks and those in charge of the refrigerators, the great majority of workers are young peasant women who live in outlying villages of the Zamora valley and the region.

The three hundred women workers interviewed were selected at random from each of the six plants. On the average, between 5 and 10 % of the total female workers in each plant were interviewed, with the exception of El Duero, where 18.3 % of the women were surveyed. Interviewed in proportional numbers, they perform the different tasks described next:

1. Stem removers—Women who remove the stems of the strawberries do piecework, that is, they are paid $5 (US $.23) per crate of strawberries, each weighing 7 kg. A worker with magic hands is able to remove the stems of up to thirty-five crates of strawberries per day; one with slow hands can barely manage five crates per day. But the number of crates available to work on varies from week to week. For example, on 4 February 1980, the four hundred workers at Frutas Refrigeradas were assigned only one crate of strawberries each, because it had rained the previous week and very few strawberries had been harvested. On days like this the expenses of the workers for transportation and food are the same, but they earn only according to the number of crates they finish. On average, 80 % of the women workers in the plants do this type of work; in the sample taken for the survey, they represent 75 % of those interviewed.

2. Supervisor—These women are chosen by the head of personnel, or by the union leader, to check whether the strawberries tossed into the canals have had (he stem properly removed. They represent 4 % of those interviewed, which is equivalent to the proportion of women working as supervisors in most plants.

3. Selectors—Once the stem is removed, strawberries float along canals filled with water and disinfectant until they reach the conveyor belts, where the selectors pick out defective or rotten strawberries. As in the case of the supervisors, the selectors are chosen by the head of personnel or by the union leader, both of whom frequently show favouritism toward their friends or toward women from their own villages. This type of work is done by about 15 % of the women workers in the plants and by 18 % of those surveyed.

4. Tray workers—From the conveyor belts the strawberries are put in tins or small boxes to be frozen, the best being placed on trays and frozen individually. This work is done by women who are selected in the manner described for those performing the two previous tasks. The women who performed the last three tasks mentioned were paid hourly, at the rate of $14 (US $.66) per hour during the 1979–1980 cycle. Though a stern remover who works with amazing dexterity might earn a higher wage than women engaged in the other tasks, normally supervisors, selectors, and tray workers earn more. Those who work on an hourly wage enjoy greater prestige because they earn more and are closer to the higher-level employees. Many of the stem removers would prefer to work on the conveyor belts, especially those who, because of their age, are no longer able to work at high speed. But seniority normally is not taken into account for either promotion in tasks or other fringe benefits. The younger workers sometimes resent the *favouritism*, not so much *for* personal *reasons*, but *rather because* of loyalty to their villages:

See here—why aren't *there more from Tinguincharo* on the conveyor belts?

But others say that it is a tiring job. For instance, Berta Olivares prefers working as a stem remover, because,

> We can at least go and walk around a little when we go get a crate for strawberries to de-stem…but those on the conveyor belts are damned uncomfortable, they don't even let them move, they can hardly even sigh. We can even sing.

Now that the scene of the work has been described, the first questions to be answered are: *Who* are these women? Did *they work before*? *If* so, what jobs did they hold?

7.6 Occupational Background of the Workers

Of the women surveyed, 61 % stated that they have never worked before. It must be noted that these included those who, because they are very young, had not yet entered the work force. Those who had worked before going into packing and freezing plants (41.3 %) performed the types of jobs indicated in Tables 7.1 and 7.2. More than half worked in agriculture previously, and a third passed through paid domestic service. Their agricultural wage labour in the region has been replaced by immigrant labour, but this is not the case in paid domestic work, since housewives in Zamora repeatedly complained that "you just can't find servants around here anymore."

Table 7.2 shows that of the formerly wage-earning women, who the strawberry agroindustry has attracted, most have been servants and agricultural labourers. We can now ask, *why* have they taken jobs in the straw berry plants? Most of the female employees prefer to work in these plants rather than as servants because, Irma Cortes said,

We are not subject to the will of '*la patrona*' (the employer) and we can live in our own homes in the village where we have friends.

Some of them like working in agriculture, but they find the work harsh. One of them said she preferred work in the fields

> …because we are out in the air, and not under the discipline of the factory, even though it is much more tiring work; for example, pulling out the weeds growing in the fields is awful hard work, and one ends up with one's back real tired.

Did they change jobs because of wage differentials? The income of 76.0 % of those interviewed increased with their employment in the plants, while that of 7.6 % remained the same. The high percentage of those who earned lower incomes (16.6 %) can be partly explained by the fact that many of these had only recently joined the plants and had not yet acquired the necessary skills, while others attended work irregularly. Of those who previously held jobs, 66.1 % worked in their own community, 28.2 % in the region, and only 0.7 % in another state, in Mexico City, or in the United States. Clearly, the strawberry companies have not brought back women working outside the region, nor have they attracted migrants from outlying regions, for the recruitment system precludes doing so. In fact, only 6.7 % of the female workers were born in another state, and more than half (60 %)

Table 7.1   Workers' previous employment by sector

Sector	Cases	
	(N)	Percentage
Agriculture	69	55.7
Services	38	30.7
Industry	7	5.8
Trade	7	5.8
Agroindustry	2	1.7
Handicrafts	1	0.3
Total	124	100

Table 7.2   Workers' previous occupations

Position	Cases	
	(N)	Percentage
Agricultural labourer	52	41.9
Servant	25	20.2
Unpaid family worker in agriculture	17	13.7
Office or shop employee	13	10.5
Factory worker	11	8.9
Trader	2	1.6
Others	4	3.2
Total	124	100

were born in Zamora and Jacona, or in Ecuandureo, a neighbouring municipality. The rest come from other municipalities in the same region.

None of the women workers live by themselves or with friends. With one exception—a woman who was adopted by I the family with whom she lives—they all live with family or kin. The fact that they still live with their families is due to a very deeply rooted social rule that forbids a young woman leaving her father's home except through marriage. But their choice of residence is also directly enforced by the acute housing scarcity in Zamora and Jacona and by the fact that the wages they earn are insufficient to permit living in a boarding-house, the only socially acceptable form of habitation for single women living away from home.

7.7 Age, Marriage and Schooling

Most of the workers (>8.7 % of the sample, are between 15 and 24 years of age (Table 7.3). Managers of the plants stated that they prefer to hire young women because of their higher productivity, and because they are 'very quick with their hands' and 'concentrate belter than the men.' In fact, the younger women's manual

Table 7.3 Age of female workers

Age	Cases	
	(N)	Percentage
12–14	30	10.0
15–19	141	47.1
20–24	65	21.6
25–29	16	5.3
39–50	39	13.0
51–80	9	3.0
124	300	100

dexterity is crucial in the task of removing the stems, but it is of only secondary importance in selecting and packing the strawberries; older women could do the latter tasks just as well. In only two of the plants, however, were older women pre-dominantly chosen for these. Additional factors that influence the preference for hiring-young women are analysed in the next sections.

Girls usually begin to work in the plants when they are 12–15 years old, and they work until they marry, normally between the ages of 17 and 21. As one of them put it,

The women marry before they are twenty because at that age the men say we have already missed the last boat.

Those who do not marry continue working, and a few young married women return to work in the plants.

Of the female workers interviewed, 85.3 % are single, 9.0 % are married, 3.0 % are divorced or abandoned, and 2.7 % are widows. Almost all workers over the age of thirty are widowed, divorced, or separated from their husbands. Most of them support their children, and perhaps their parents or siblings as well. The few married women workers state that their husbands do not send back enough money from the United States, where they are working.

One older woman told us that in the early times of the packing plants women stood in long lines outside of the plants hoping to be hired:

There were little girls, young girls and adults, even old women.

But at present, the increase in the number of plants has led to a relative scarcity of women workers, particularly during die peak time of the season. At this busy time, plants hire women of all ages, including twelve-year-olds and older women. Then, as strawberries begin to come in at a slower rate, the management begins to eliminate workers "First the little girls, then the lazy ones, then others begin to drop out by themselves when I hey see that there is very little working time left", one worker told us.

Sixteen percent, mostly the older workers have not been to school at all, while 31 % attended primary school up to the third grade. This low average in schooling can be explained by conditions in their communities, but it is sig-nificant to note that 3.7 % have reached the high-school or preparatory-school level, since in theory their education should have given them access to jobs with

Table 7.4   Age of women workers in Aguascalientes and Zamora

Age	Cases	
	In Zamora (%)	In Aguascalientes (%)
12–19	57.0	52
20–29	27.0	21
30–39	7.7	10
40 or over	8.3	17
Total	100	100

Sources For Aguascalientes: Díaz Ronner/Muñoz (1978). Other statistics from authors' research

Table 7.5   Schooling of women workers in Aguascalientes and Zamora

Schooling	Cases	
	In Zamora (%)	In Aguascalientes (%)
None	16.0	32
1st–3rd grade, primary	31.0	28
4th–6th grade, primary	49.3	40
Secondary or preparatory	3.7	–
Total	100	100

Sources For Aguascalientes: Díaz Ronner/Muñoz (1978). Other statistics from authors' research

higher incomes and prestige, But the fact is that very few such jobs are available in Zamora, and, besides, these women explained that they earn more money working at a fast pace in the plants than they would working in a shop or an office.

Although they seem to recognize this, the great majority of the women are convinced that their low degree of schooling prevents them from getting other jobs, and they complain bitterly that their parents, especially their fathers, did not allow them to go on studying "Women are not allowed to finish [school] because our parents say it does not pay for itself because we then go and get married, and it has only been a waste". Another one said "If I were to study, I could be a secretary, and I would stop doing this very tiring job". The mythical nature of this hope becomes clear if we realize that, as has happened in other developing countries, an increase in levels of education would lead to an increase in job entrance requirements, and consequently, the same proportion of less qualified women—even if their educational level were higher in absolute terms—would continue filling the lower-level jobs.

This hypothesis is further strengthened if we compare the plant workers surveyed in Zamora with a group of female agricultural labourers, surveyed in the state of Aguascalientes north of Michoacán, who pick grapes seasonally. The profile of marital status among the grape pickers resembles that of workers in the strawberry plants: 80 % are single, 8 % are married, 3 % are divorced or separated, and 9 % are widowed (Díaz Ronner/Muñoz 1978: 331). In ages and schooling, the percentage distribution is also similar, but there are significant differences.

The similarities in both age structure and schooling indicate that roughly the same social groups of women enter either of those jobs (Tables 7.4, 7.5).

But more women with higher schooling between the ages of twelve and nineteen enter strawberry-factory work in Zamora. The foregoing suggests (1) that many young, single girls enter agroindustry who otherwise would not work for wages and (2) that strawberry-plant work attracts women whose higher educational levels make it unlikely that they would accept work in agriculture. However, additional data not included in the surveys on the educational levels in the communities would be necessary to confirm the latter hypothesis.

7.8 Social Attitude Toward Women's Work in the Factories

When the strawberry industry first began, it was very difficult for the plant managers to recruit enough women workers. They could get those who were already working in other jobs but were unable to attract young women whose families were not in dire need of additional income. The women's reluctance to enter paid employment was due to the very real fear, confirmed by women's experiences, that unaccompanied young women in public places would be 'stolen.' Carmen García summarises it neatly;

> Previously, it was really rotten for the girls, because they were frequently stolen when **they** were going to fetch water, or to wash clothes or to bring the '*nixtamal*' (**maize dough**)... they were even stolen with the help of a gun or a machete. They were taken into the woods and then the men would come to ask their parents for the girl (in marriage). Most of the girls did marry them, even if they did not want to, and here divorce is out of the question. If they don't get on together, the woman just puts up with it. Here it is customary tor the husbands to beat the women when they are drunk, they say that blows make women love them more.
>
> Yet, as it happens, the fact that the young girls are no longer 'stolen' as often in the peasant villages of the region as they were in the past is attributed mainly to their working in the strawberry plants, although no one ever explains exactly why this is so.
>
> At first, the fathers flatly refused their daughters permission to work in the factories. One woman told us "The parents are not used to one's working and in the village people gossip a lot, they say that the women who go out to work go with many men". Not long ago it was still forbidden for men and women in the villages to address one another on the street. What the parents most feared did occasionally happen. An experienced worker, Inés Gómez:
>
> When it [work in the plants] began, it turned out that many of the girls got pregnant because they did not know how to look out for themselves, and we move in an environ-ment of '*machismo*' and paternalism, it happened frequently...but now the girls know how to handle themselves, now they even want to study and improve themselves.
>
> The young women workers see their situation in a different way and complain bitterly:
>
> All they do is spread rumours about us. Many boys say they won't marry those who work in the plants, and all the girls from the village work there, but of course later they themselves are after us. They spread many untrue stories about us. Some of our nieces even went around saying that we were pregnant, and that we had left the children at the social security.

The young women workers' situation is further complicated by the migration of most of the marriageable men:

> The girls don't, go north [to the United States] because people talk badly about them. Even if we just go to Zamora they talk badly, we can never go anywhere... The boys are allowed to go north and they come back real proud, some of them shack up with the

American girls over there. They say they are very loose, that they even go after the boys. Others do return here to get married.

Initially the local priests were opposed to the women's factory work, too. One incident illustrates the situation very clearly. The strawberry plants in Jacona were unable to get female workers because every Sunday the local priest thundered that women would go to hell if they sinned by going out to work in the factories, It is said that the problem was solved when the owners of the plant spoke with the priest and offered to pay for the cost of a new altar tor the parish church. Since that day, the local priest has exalted the dignity of work.

7.9 Wages and Expenditures

As has been noted, the workers' wages are subject to the rate at which the plants buy strawberries dining the year and to their own level of skill. The monthly average wage among workers surveyed is $1,126 (US $51.18). Eleven percent earns an average of $1,750 (US $79.51) per month, 26 % earn an average of $750 (US $34.09), and 8 % earn an average of $350 (US $15.90). These wage levels are very far below the legal minimum wage, which amounted to $4,260 (US $193.63) for the region in 1980. Since a single person, let alone a family with children, cannot survive on tins income, such low wages can only be considered as complementary to the main income of a family.

Worse still is the fact that the wages these women get vary enormously on a day-by-day and week-by-week basis. The season begins in November or December and lasts until July or August. However, during that period there are 'bad months' as the women call them—November, December, January, February, August, September—in which they earn an average of less than $500 (US $22.72) per month. During the good months they may earn as much as $2,200 (US $90.90) per month. Most of the women are not hired at the plants for the whole year; 56 % work from 7 to 9 months; 5 % work from 10 to 12 months; 16 % from 4 to 6 months; 11.6 % from 1 to 3 months; and 11.3 % do not get to work even 1 month per year. Many of those in the last group work only on Saturdays during the peak season, or they are younger sisters of the workers, and tag along a few days per week.

During the months when there is no work in the plants, 75.3 % remain at home helping with the domestic work; some do embroidering or knit pieces for sale. The surprisingly large number of women who follow this pattern indicates that these families do not urgently require a constant income from the women workers. In some cases—as, for example, one where the daughter supports herself and her mother—the income earned in the plant in the months of seasonal work is sufficient to keep them during the 3 months without work. Among the 24.3 % of the workers who do work during these months, 7 % work as servants, 11 % go harvesting in the fields as day labourers, 1.0 % work in offices, and 0.3 % migrates to the United States. The remainder works in the informal sector in a variety of ways.

To what extent does these predominantly peasant families depend on the women workers' income? The majority (61.6 %) answered that their work only

Table 7.6 Proportion of workers' wages given to parents

Monthly wages (pesos)	All	Almost all	One half	A little	Nothing
200–1,000	36.9[a]	15.7	23.8	10.5	13.1
1,001–2,000	30.1	25.3	27.4	12.4	4.8
2,001–3,000	38.2	27.4	25.4	7.2	1.8
Over 3,000	11.2	44.4	11.1	22.2	11.1
All wage categories[b]	31.0	24.0	25.7	11.0	5.7

[a]Numbers given in percentages
[b]2.6 % of workers surveyed did not answer this question

partially supports their families, 20.7 % replied that they give no financial help to their families, and 17.7 % stated they offer major support. It is usual for one of the younger girls to hand over the entire weekly wage to her father or mother, who then little by little lets her have whatever money she requires for her expenses. Table 7.6 shows that the correlation between the amount a worker gives her parents and the amount she earns is not significant.

How are their wages spent? What the workers keep for themselves, they spend on fashionable clothes, costume jewellery, romantic comics and stories, and beauty products. But the larger part of their wage, handled by their parents, goes into buying household consumer goods. This has helped the shops selling furniture and electric appliances. Some of the consumer goods purchased in the poorer households are basic items such as gas stoves, beds, wardrobes, and sewing machines; in other households the goods may be televisions, radios, blenders, and record players. Only a few households buy luxury items such as enormous consoles, fancy furniture, porcelain figurines, wine glasses, and so on. The survey indicates, however, that the parents buy these items not only tor prestige but also because they can sell or pawn them when times get hard. It must be noted that the commercial boom in Zamora is due only in part to the women workers' income; it is mainly a result of the income in dollars sent back by the male migrants working in the United States. Even so, the pattern of consumption is the same in both cases.

7.10 Recruitment of Workers for the Plants

Women workers are recruited each season through social networks in the communities. In the plants that have unions, the union secretary chooses women delegates in each village or hamlet; in plants that don't have a union, the head of personnel chooses these delegates. Once the word is sent to them that they should begin recruiting, these delegates go around the village letting everyone know that they are hiring. They list the names of those women who want to go to work, purportedly giving preference to experienced workers. But Antonieta Castro complained that previous experience matters little "Some of the new ones are given preference by the bosses, because they give them '*gollete*' (some present). We don't get angry about this, we only feel hurt". The 'loyalty' that a worker has shown toward the general secretary of the union of the company is also taken into account during compilation of the lists, as are personal preferences and group rivalries within the Community. In hiring, the company follows the list made by the delegate, moving through it progressively as the season advances. The recruitment in the village, usually an older woman, is socially responsible for the young girls she recruits as workers. Parents sometimes allow their daughters to go only lf they trust the recruiter. This responsibility also gives the **latter** the power to decide who will work in the plant.

7.11 Conditions of Work

Hiring conditions and *benefits* in most plants are clearly below legal requirements stated in Mexican law. In the first place there are no contracts or permanent jobs for the workers. According to the law the companies should pay the minimum wage, establish fixed working schedules, and hire the workers permanently during the entire year. In the second place, fringe benefits are non-existent: plant workers have no Social Security, nor do they have adequate medical services. More mothers could work if the plants had nurseries, and by law factories must provide one whenever there are more than thirty permanent women workers. When the women ask for a nursery, however, they are turned down. One manager said "We saw that the nursery was not really necessary because only two or three children come along with their mothers, and that is why we did not put one in".

7.12 Women's Perception of Their Work in the Plants

Although these conditions persist, and in spite of the fact that many of the women employed in the plant consider the work to be tiring and oppressive, they prefer it because their only alternative is to remain shut in their homes doing domestic work or to work in jobs that are even more underpaid. Of the workers surveyed 65 % said that they prefer to work outside their home. As Amalia Vega put it:

> We like so much to go out and work in the packing plant that when we return to our village in the evening we skip along the road dancing and singing, We don't mind about being tired [after a working day of eight to eleven hours]; because we have earned *our* few pennies and have left the little ranch for a while, we are very happy. In the village you get bored by seeing the same faces all day long and listening to the same gossip. By working we entertain ourselves.

This is, in fact, a very fair assessment of the situation. When asked what type of work they like best, 59 % answered that they prefer to work in a strawberry plant;

only 4.5 % prefer to work on the land, and 36.5 % would prefer to be employed in an office.

Although four out of every five workers interviewed said they wanted improvements in their working conditions, particularly in wages and in the treatment they receive from their bosses, there are no real channels for protest. Only half of them belonged to a union, but this was due to the fact that only four of the six plants had a union. However, less than half of the workers (46.7 %) thought that unionization could help them get better working conditions. This distrust reflects the fact that the existing unions closely collaborate with management. The pragmatic attitude of the union leaders, some of them women, is evident in the statement of one woman leader. Asked how she and other leaders got along with management, she said:

> Fortunately there has always been a good relationship. People get to understand each other by talking. Also, we are interested in the company not having a loss, otherwise, we don't get 'utilidades' (a profit-sharing government scheme).

In actual fact, workers rarely receive 'utilidades', which are sometimes used to pay for the annual fiesta and mass in the plant. As a result, workers hardly participate in union activities: "We get bored going to the meetings", one worker told us. "We don't understand anything and we get nothing out of it. We just waste our time."

Almost all the younger workers consider their job in the agroindustry as a stage in their life that allows them to get out of the daily routine of the village. More than half (58 %) answered that they do not plan to go on working once they get married. As one of them told us, "Why, that's what I'm getting married for, to stop working". Of those who say they may continue to work, many said they might marry a 'bum' and end up having to support their household.

7.13 Conclusion

Why does the strawberry agroindustry predominantly employ women? It is true, that the jobs of removing stems and selecting strawberries require manual dexterity but so do surgery, dentistry, playing musical instruments, and other predominantly male occupations. There are, then, other reasons why this industry employs women. In the region of Zamora, agroindustry cannot compete with the wages paid in the United States in order to attract and retain migrant male labour. At the same time there is a large, population of young women who have very few alternatives for work. The strawberry plants do not have to compete with urban wages for women workers, since the emigration of women from the region is not frequent; male emigration largely covers the deficit in the budget of most peasant families. Moreover, the great majority of young women in peasant families have access only to paid domestic work or to wage labour on the land, both of them unrewarding jobs.

Therefore, a major reason for employing women is that they can be paid much lower wages than those stipulated by law; and can be asked to accept conditions in which there is a constant fluctuation in schedules and days of work. Here it seems to us that the companies take advantage of the traditional idea that, any income earned by a daughter, wife, or mother is an 'extra' over and above the main income of the father, husband, or son. If such wages were paid to male workers, the low income and the instability of the job would be untenable in the long run; workers would either move to other jobs or organize and strike for higher wages.

Other results of the analysis support this view. The fact that the percentage of women household-heads in the packing and refrigerating plants is very low— 5.7 % as compared to 12 % in the legion as a whole—suggests that the wages paid by the plants cannot constitute the central income of a household. Of course, it also reflects the factories' preference for young, unmarried workers.

Thus, the plants attract many young women—approximately one-half of the women workers—who normally would not enter wage labour if the plants did not exist. Or so it seems, at least, from a comparison made with a group of agricultural labourers from Aguascalientes and from the fact that 42.4 % of the workers surveyed gave only half or less than half of their wages to their households. Further support for this hypothesis is found in the large majority of women workers who do not seek alternative work during the months they are not employed in the plants.

Another advantage for the plants is the constant turnover among women workers. This impermanence allows a company considerable savings in wage increases due to seniority as well as in payments for maternity, illness, or disablement and in old-age pensions. Importantly it also prevents the workers from accumulating information and experience that would lead them to organize and to demand improvements in hiring and working conditions. Mean while, the traditional culture itself reinforces this turnover by making marriage the only aspiration for women.

Clearly, the strawberry agroindustry in Zamora is very profitable because cheap female labour is readily available. This conclusion coincides with that reached by Feder (1977), who points to the low cost of labour as one of the most important factors in making the Mexican strawberry industry competitive internationally. Thus the 'comparative advantages' of this industry in the international market are closely associated with the 'comparative disadvantages' of young inexperienced, rural women who face social, legal and economic discrimination. From a sociological point of view, what the agribusiness capitals have done is to make use of certain social and cultural characteristics of the region, that is, the high demographic growth, the traditional cultural values that assign a subordinate role to women, the family structure of the communities, and the local patterns of consumption. The key question to be asked is whether this way of using resources will improve the living conditions of the women and of their communities.

Have conditions of life for these women changed with their entry in salaried industrial work? This study shows that they have changed very little. The great majority of workers continue to live in their parents' homes; very few have gone to live with other relatives in Zamora, and always under the same conditions of subordination and restriction they experienced in their own homes. About half of them hand over the greater part of their wages to their parents or use their earnings to support their own families. Thus they have only slightly increased their personal consumption. Their families, of course, have an improved standard of living, at least temporarily.

Although the women have more freedom when working outside of their homes, they are harassed by the men in the streets and are not free to move around the town or the villages on their own. Even when traveling to and from the plants the young women are closely supervised by recruiters and union leaders. There have been some changes: the young girls are not 'stolen' as frequently as before and apparently they have a more decisive voice as to whom they will marry. Also, some have become eager to study and to get ahead.

However, for the majority of women, work in this agroindustry is no way to get ahead. There are no promotions; the workers get no encouragement or help in acquiring skills or education; and the instability and low wages of the jobs do not offer any prospects for improvement in the future. Predictably, under these conditions no significant cultural change is taking place. On the contrary, the lack of prospects or promotion in the agroindustry, the low wages and the high level unemployment only push the workers back into the traditional hope of marriage as the only road toward a better future. Only a few of the young women, mostly those who have not married, have acquired new aspirations about employment possibilities and life styles. For them, however, it will be very difficult to find employment once the strawberry industry declines. The fact that the strawberry companies take advantage of the traditional values and conditions that subordinate women means they have an interest in reinforcing this traditional order. In fact, it is in their interest to oppose any initiatives to change the passive, submissive role of women in Zamora. In this sense, no 'modernization' of women's roles is evident in the region.

What has been the impact of the strawberry agroindustry on the communities of the region? In the short run the industry has provided a better standard of living for rural families. The majority use the women's incomes to improve their housing and, particularly, to purchase household goods—furniture and electric appliances which also serve as a form of saving. The worker's wages, then, flow rapidly through the merchants of Zamora toward the urban industries that manufacture these consumer goods.

But while the market for consumer goods has expanded, the *poorer* groups in the region have not been brought into the market. Because of the hiring practices in the plants, work is not given to women heads of household, nor to the poorer male and female labourers—those who most require an income. Rather, since the survey shows that the majority of the workers do not support themselves, it suggests that jobs are given mostly to young women from the mid-level peasant families, whose wages serve to improve their families' standard of living. Although such a gain is not to be underestimated, it benefits only minimally those households whose economic survival depends entirely or partially on women's wages. As a result, older female heads of household are pushed back into the strenuous, harsh, and even more poorly paid job of strawberry picking in the fields.

The strawberry agroindustry is not creating conditions for the future development of the region. It is not training workers, nor is it promoting or improving social services. It has not stopped the emigration of men to the United States. Finally, it has tended toward the concentration of land and capital while displacing and undermining production in small landholdings.

Thus, it seems to us that the strawberry agroindustry has provided some short-term improvements, but in the long run—aside from the profits that flow mainly to US agribusiness concerns and to affluent local entrepreneurs—it will leave behind nothing but ashes when it collapses. According to two plant managers, the collapse is expected in 3–5 years. It is difficult to refrain from apocalyptic forecasts when we can see that the decline of this agroindustry will plunge the region back into underdevelopment; peasant household incomes will fall, massive unemployment will force countless women and families to migrate, and the hopes for a better life that have been raised among women will, once gam, be destroyed. Basically, nothing will have changed for women. Since the strawberry industry requires female workers whose income is not essential for the household, it bypasses the needy and predominantly employs women from middle-income groups. Since it requires submissive and docile workers, it reinforces patriarchal and authoritarian structures. Since it benefits from a constant turnover of workers, it does not oppose the *machismo* that confines women to home and marriage.

Two dilemmas emerge from a feminist analysis. The first is that much of the data—for instance, Amalia Vega's touching description of the joy she and other women feel at being allowed to leave the narrow horizons of their villages—show that the plants improve the lives of women and therefore, from a feminist point of view, should be defended. At the same time, salaries and working conditions at these plants are dismally exploitative, comparing unfavourably both to the norms set down by Mexican law and to actual situations elsewhere in Mexican industry; for this reason they should be denounced and opposed.

An even more painful dilemma faced by women's movements in situations such as this is that women whose consciousness has been raised by temporary employment will be left stranded when economic and social survival again becomes difficult if not impossible, when such industries move to regions populated by another group of docile and disadvantaged women. Thus, by the time the strawberry agribusiness—or the US assembly plants along the Mexican border, for that matter—move to other countries that offer still lower production costs, the jobs Mexican women had temporarily gained from the loss experienced by their US counterparts will also be lost to them. The jobs will then become a temporary gain for, perhaps, Haitian or Honduran women.

In this way, women's 'comparative disadvantages' in the labour market in any given country can, at some time, be translated into 'comparative advantages' for companies, capitals, and governments in the international markets, but when disadvantaged women organize to get even minimal improvements in wages and working conditions, the 'comparative advantages' are lost, and investment goes elsewhere. Clearly, all women lose along this chain. This being the case, one can only conclude that discrimination against women in employment, reflecting as it does the disadvantages women suffer from attitudes about gender, from social customs, and from their lack of political power, cannot be fought effectively in one place or country unless an appropriate international perspective is developed.

References

Arizpe, Lourdes, 1978: *Migración, etnicismo y cambio económico* (Mexico City: El Colegio de México).

Burbach, Roger; Flynn, Patricia, 1978: "Agribusiness targets Latin America", in: *NACLA, Report on the Americas*, XII, 1 (January–February): 2–35.

Deere, Carmen Diana; León de Leal, Magdalena, 1981: "Peasant Production, Proletarization, and the Sexual Division of Labor in the Andes", in: *Signs*, 7, 2 (Winter): 338–360.

Díaz Ronner, Lucila; Muñoz, María Elena, 1978: "La mujer asalariada en el sector agrícola" in: *América Indígena*, 38 (April–June): 327–334.

Feder, Ernest, 1977: *El imperialismo fresa* (Mexico City: Editorial Campesina).

Fernández-Kelly, María Patricia, 1983: "Mexican Border Industrialization, Female Labour Force Participation and Migration", in: Nash, June; Fernández-Kelly, Maria Patricia (Eds.): *Women, Men and the International Division of Labour* (New York: State University of New York): 205–223.

George, Susan, 1977: *How the Other Half Dies* (Montclair, NJ: Allan Held, Osmun & Co).

Hewitt de Alcántara, Cynthia, 1978: *La modernización de la agricultura mexicana* (Mexico City: Siglo XXI Editores).

ILO (International Labour Organization), 1980: *Women in the Economic Activities of the World: A Statistical Analysis* (Geneva: OIT).

León de Leal, Magdalena; Deere, Carmen Diana, 1980: *La Mujer y el Capitalismo Agrario* (Bogotá: ACEP).

Lim, Linda, 1978: "Women Workers in Multinational Corporations: The Case of the Electronics Industry in Malaysia and Singapore", Michigan Occasional Papers, No. 9 (Ann Arbor, University of Michigan).

Massiah, Joycelin, 1980: "Family Structure and the Status of Women in the Caribbean, with Particular Reference to Women Who Head Households". Paper presented at the Conference on Women, Development and Population Trends, UNESCO, Paris, November, reference UNESCO, SS-80/ Conference 627/Col. 34.

Medrano, Diana, 1980: "El caso de las obreras de los cultivos de flores de los municipios de Chia, Cajica y Tabio en la sabana de Bogotá, Colombia". Research paper for the Rural Employment Policies Branch, International Labour Organization, Geneva.

Moore Lappe, Frances; Collins, Joseph; Fowler, Cary, 1977: *Food First: Beyond the Myth of Scarcity* (Boston: Houghton Miffin Co).

Palmer, Ingrid, 1977: "Rural Poverty in Indonesia with special reference to Java", in: *Poverty and Landlessness in Rural Asia* (Geneva: International Labour Office): 205–231.

Rama, Ruth, 1978: "Empresas trasnacionales y agricultura mexicana: el caso de las procesadoras de frutas y legumbres", in: *Investigación Económica*, 37, 143, (June): 75–118.

Rendón, Teresa, 1976: "Utilización de mano de obra en la agricultura mexicana, 1940–1975", in: *Demografía y Economía*, 10, 3 (April–June): 352–385.

Roldán, Iris Martha, 1980: "Trabajo asalariado y condición de la mujer rural en un cultivo de exportación: el caso de las trabajadoras del tomate en el estado de Sinaloa, México". Research paper for the Rural Employment Policies Branch, International Labour Organization, Geneva.

Safa, Helen I., 1981: "Runaway Shops and Female Employment: The Search for Cheap Labor", in: *Signs*, 7, 2 (Winter): 418–433.

Silva de Rojas, Alicia E.; Corredor de Prieto, Consuelo, 1980: "La exportación de la mano de obra femenina en la industria de las flores un estudio de caso en Colombia". Research paper for the Rural Employment Policies Branch, International Labour Organization, Geneva.

Spindel, Cheywa R., 1980: "Capital Oligopólico y a producao rural de base familiar papel socio-economico da mulher". Research paper for the Rural Employment Policies Branch, International Labour Organization, Geneva.

Todaro, Michael, 1977: *Economics for a Developing World* (London: Longman Group Ltd).

UNIDO (United Nations Industrial Development Office), 1980: "Women in the Redeployment of Manufacturing Industry to Developing Countries", UNIDO Working Paper No. 18 (Vienna: UNIDO).

UNDP (United Nations Development Program), 1980: "Rural Women's Participation in Development", UNDP Evaluation Study, No. 3 (Geneva: UN).

Young, Kate, 1982: "The Creation of a Relative Surplus Population: A case study from Mexico", in: Beneria, Lourdes (Ed.): *Women and Development. The Sexual Division of Labor in Rural Societies* (New York: Praeger): 149–178.

Zeidenstein, Sondra, 1979: "Learning about Rural Women" in: *Studies in Family Planning*, Population Council, 10 (November–December): 1–114.

Young girl in back-breaking job of picking strawberries in the field, Zamora, 1980. This photo is from the author's photo collection

Chapter 8
Mexican Agricultural Development Policy and Its Impact on Rural Women

Lourdes Arizpe and Carlota Botey

8.1 Introduction

Over the past decade, an increasing number of studies have focused on rural women's participation in production and reproduction. Few of these studies, however, have explored the relationship between women's participation and state agrarian policy. Nevertheless, women's participation is directly affected by changes in the productive structure induced by agrarian policy. Moreover, the gender division of labor within the peasant household itself generates differential effects on agrarian processes. The topic is complex, and before assessing the impact of agrarian policy on rural women, we will attempt to organize the available data and then propose some hypotheses.

The Mexican Agrarian Reform created the conditions for development of the peasant economy by establishing the *ejido* and the Indian community as the basic units of social property. These divisions were conceived of not simply as units of agricultural production but also as social and cultural entities encompassing all sectors of the population—men, women, children, youth, and the elderly—all of which had defined places and specific functions and employment within the productive community. However, the workings of the market economy, together with demographic growth and the urban bias in public policy, have brought about the breakdown of the internal social organization of the *ejidos* and Indian communities in recent decades. This breakdown has resulted in unemployment, migration, and cultural disruption.

Carlota Botey—Deceased

This chapter was translated by Katherine Pettus. It was originally published as: "Mexican agricultural development policy and its impact on rural women", in: Deere, C. D.; Leon, M. (Eds.): *Rural women and state policy: feminist perspectives on Latin American agricultural development* (Boulder: Westview, 1987): 67–83. Permission was granted by the coeditors Carmen Diana Deere and Magdalena Leon on 27 November 2013 to whom the Perseus Books Group returned the copyright.

L. Arizpe, *Migration, Women and Social Development*,
SpringerBriefs on Pioneers in Science and Practice 11,
DOI: 10.1007/978-3-319-06572-4_8, © The Author(s) 2014

The peasant sector in particular has been affected by changes in women's economic and social participation. Women play a fundamental role in maintaining family and community cohesion *and in transmitting* basic cultural norms and social mores. In addition, they perform both primary and secondary tasks in agricultural production. In spite of the central role women play in rural community life, agrarian planners have failed to consider the various ways that women participate in farm and agro-industrial production, their access to income, and their role in social organization. The agrarian reform benefited only male heads of households and ignored the question of how economic change would affect the access of other family members to productive inputs and to income-generating activities.

8.2 The Agrarian Reform

Mexican agrarian policy, whose centerpiece is the agrarian reform—defined as the set of legally enacted economic, social, and political policies that gave land rights to the peasantry—reflects the historic struggle between collective and private forms of landholdings (Reyes Osorio 1972). The 1915 Agrarian Reform legislation in Mexico, promulgated after the 1910 revolution, was the first of its kind in Latin America. The reform replaced the Roman legal principle of unrestricted private property with one that recognized, first and foremost, that land and water rights must serve a social function. Over six decades, the obsolete *latifundio* system that had dominated the Mexican countryside for more than four centuries was dismantled. Gradually, a more democratic land tenancy system, which encompassed both smallholder private property and the social property of the *ejido* and Indian community structures, was consolidated.

This social property combined indigenous, Spanish, and English forms of land tenancy: it reestablished the pre-Columbian communal form of property enriched by European colonial influence as well as the principles of the North-American homesteading movement which, in the Mexican Civil Code, has its counterpart in the family holding.

As forms of landholding, the Mexican *ejido* and Indian community system are based on the following principles:

1. Land, water, and forests returned or given to an *ejido* or recognized as belonging to a community are the property of each village or agrarian unit granted legal status by presidential resolution; land is not nationalized.
2. Under the Mexican Civil Code, *ejido* and communal lands have the same legal characteristics as family holdings, being inalienable private property.

3. Within each agrarian unit, the members included in an official census have individual rights and obligations with regard to the personal usufruct of the *ejido*'s resources, rights that can be inherited by economic dependents in accordance with the relevant law.

Land redistribution began in 1915 but did not gather momentum until 1934 when President Lázaro Cárdenas' administration foresaw the development implications of collectively held property and community organization. Between 1915 and 1983, approximately 103 million hectares were redistributed to 25,589 *ejidos* and 1,486 Indian communities, benefiting 2.8 million household heads.

Although almost half of the land in Mexico is now collectively owned, its development potential is limited because only 11.8 % of this total area is suitable for cultivation and only an additional 2.2 % has access to irrigation. The bulk is largely grazing land of varying quality; although some is wooded, much of the area is either unsuitable for commercial agriculture or livestock or is useless. Nevertheless, the successful growth (more than 5 % annually in the 1930s and 1940s) of agricultural production on agrarian reform lands provided the cheap food, labor power, and resources Mexico needed to industrialize in the 1950s.

The agrarian reform, besides establishing *ejidos* and titling community-held land, consolidated a sector of individual private property through the subdivision of the *latifundio* and the titling of national lands. There were fewer than 50,000 smallholders in 1910; they now number over one million.

Although rural living standards are clearly higher than they were seventy years ago, peasant income levels are still quite low, particularly relative to those of other social groups. Land redistribution must be recognized as only a precondition for a more extensive program of rural development. However, the economic and social efficiency of the peasant sector and of social property was greatly undermined in the 1950s and 1960s when state investments favored capitalist private property.

The new agrarian problem centers on the uneasy balance between the two main agricultural sectors, one based on social property and the other on capitalist private property. To this problem is added that of the large mass of landless agricultural laborers who compete with smallholders for seasonal wage work in the countryside. Once rural population growth exceeded the physical capacity of the distribution program, the problem of insufficient employment generation in manufacturing could no longer be resolved through the periodic redistribution of land. Yet no viable alternatives for the landless have been proposed.

The contemporary agrarian problem demands long-range planning of new productive activities and the requisite training programs.

LOURDES ARIZPE

LA MUJER EN
EL DESARROLLO DE MEXICO
Y DE AMERICA LATINA

UNIVERSIDAD NACIONAL AUTONOMA DE MEXICO
CENTRO REGIONAL DE
INVESTIGACIONES MULTIDISCIPLINARIAS

Book cover of *Women and Development in Latin America* (Mexico City: Centro Regional de Investigaciones Multidisciplinarias-National Autonomous University of Mexico, 1990). The book title is reproduced with the permission of CRIM-UNAM

8.3 Women's Rights in the Mexican Agrarian System

The first agrarian law to be promulgated after the Mexican Revolution (January 6, 1915) made no specific reference either to individual land rights or to the size of landholdings beneficiaries were to receive: land was either given or returned, with legal title, to communities. The land rights clauses in the 1917 constitution also made no reference to gender (Chávez 1974).

The 1920 *Ejido* Law—the first piece of legislation to establish that land should be distributed equitably among heads of households—made no mention of women. Article 9 of the By-Laws, ratified in 1922, stated that "wherever land is granted to *ejidos*, the heads of households or individuals over the age of eighteen shall receive from three to five hectares of irrigated or rainfed lands." The 1927 law is the first statute to refer to women: Article 97 establishes that *ejido* members shall be "Mexican nationals, males over the age of eighteen, or single women or widows supporting a family." It is interesting that the 'supporting a family' qualification refers only to women.

Legal equality between men and women is not explicitly established until 1971—nearly fifty years later—in Article 200 of the Federal Law of Agrarian Reform. To receive land beneficiaries had to be "Mexican by birth, male or female over sixteen years of age, or of any age if with dependents." Female *ejido* members (*ejidatarias*) were to have the same rights as males (*ejidatarios*). To stress this point, Article 45 stipulates that "women shall enjoy all the rights pertaining to *ejido* members, shall have voice and vote in the General Assemblies, and shall be eligible for all positions in the Committees and Vigilance Counsels." Articles 76 and 78 were also designed to favor *ejidatarias*:

Land rights conferred in the foregoing Article may not be the subject of sharecropping, leasing, or third party contracts, nor may land be farmed by wage workers, unless the beneficiary is a female head of household prevented by domestic obligations and the care of young and dependent children from farming the land herself, and the beneficiary resides in the community.

Individuals may not accumulate land units. If an *ejidatario* marries or becomes the common law spouse of an *ejidataria* having land rights, each shall retain their separate holding. Under agrarian law, it shall be considered that the marriage was celebrated under the rule of separation of properties.

8.4 The Women's Agro-Industrial Units

A novel feature of the 1971 law was its provision for the creation of agro-industrial units for Women (*Unidad Agricola Industrial de la Mujer*—UAIM) in the *ejidos*. All women over the age of sixteen who were not *ejidatarias* in their own right were given access to a collectively held plot of land for special agricultural or agro-industrial projects. The UAIM is, to date, the main state initiative with respect to rural women.

In the 14 years since the program's inception, 8,000 UAIMs were provided with legal status. Less than one-fourth (1,224) of this original number actually began operations, however, and credit was extended to only 1,112 units. In some cases, particularly in *ejidos* with insufficient land for the men, the latter denied women their right to a collective plot of land. In many cases, once they received it, women simply used their land to cultivate maize or other agricultural products. The majority of the projects that received credit were maize-grinding or tortilla shops, poultry and pig farms, sewing and embroidery collectives, or food canning and preserving operations. They function as production cooperatives in which the women's remuneration is determined according to the income from sales.

UAIMs all over Mexico face the same problem: despite the women's dedication, these small enterprises are not commercially viable. Part of the problem arises from the monopoly and oligopoly control of distribution outlets; as a consequence the UAIMs often cannot find a market for their products. But the quality of their products, especially garments, is also a problem. Rarely are the UAIMs competitive with the large national or transnational manufacturers because their products are more expensive than their more technically sophisticated counterparts (Barbieri 1981).

On the other hand, the UAIMs have succeeded in spreading awareness of the need to expand employment opportunities for rural women, particularly young women who prefer to stay in the rural communities rather than migrate to the cities as did their elder female relatives. The UAIMs give visibility to the potential of these young women workers. In addition, the UAIMs provide the legal context to make women's participation in *ejido* and community activities politically and socially acceptable.

Despite the provisions of the 1971 law, cultural conditioning and discriminatory patriarchal practices continue to limit women's access to land. Female *ejido* or community members accounted for 15 % of the total in 1984; the majority were elderly widows who inherited their husbands' land plots. Few participate directly in the productive process, and the control of the parcel is usually in the hands of a son or brother.

8.5 Rural Women's Program

Although the UAIM provided an institutional base for women to participate in rural production, the first large-scale program to target rural women, the 'Women's Program for Rural Development' (*Programa de la Mujer para la Consecución del Desarrollo Rural*–PROMUDER)[1] at the Ministry for Agrarian Reform, was not initiated until 1983. The program was based on the assumption that rural women's participation varied at the generational level and that women of different age groups required different sets of income-generating activities.

Accordingly, PROMUDER proposed the following projects: (1) The UAIMs were to target young women living in the *ejido* or community; (2) horticultural

---

[1] Lourdes Arizpe helped develop this program, which was set up by former Undersecretary for Agrarian Organization Beatriz Paredes; it was coordinated by Margarita Velázquez.

projects would be developed for more senior women and mothers, who had difficulty working outside their household compound; (3) women farmers would receive training in agricultural production, new technologies, and food storage and preservation; (4) younger women engaged in agro-industrial or assembly line work would receive training in manufacturing technical skills and in labor legislation; (5) poor rural women and young women working in domestic service in the cities would receive training and legal counseling in family and land rights issues; (6) all rural women were targeted for a literacy program.

The program supplied only broad guidelines, the idea being that the women themselves, in the different regions and communities, would define the details and implement the projects. Community representatives, local and state officials, and extension workers in charge of the program worked together to accomplish these objectives.

Perhaps the most striking feature of the program, as projects got under way, was the enthusiastic response it received from rural women. Although initially wary, the women were soon demanding more and more projects for their communities. The desperate need for cash income, combined with the lack of employment opportunities for women in the communities and their new-found awareness that they could participate fully in economic activities, provided a strong incentive to experiment with the various projects.

Inevitably, however, problems arose. In some communities, the men were either uninterested in helping to set up the women's projects or openly hostile to them. Having little experience in working as a team, the women had difficulty in getting organized, as well as in carrying out accounting, quality control, and marketing operations. This lack of organization and training sometimes resulted in the extension worker's becoming the dominant figure in the project, and the women's participation and opportunity to learn new skills were curtailed. Nevertheless, the very fact that such projects were proposed at the upper policymaking echelons and that they were initiated throughout the country with a ripple effect meant that rural women found new avenues for participation open to them.[2]

8.6 The Agrarian Crisis

Mexican industrialization began in the 1940s, stimulated by the US wartime demand for manufactured goods, and continued as an import substitution program under a policy of 'stabilizing development' until the end of the 1960s. Under this economic development model, the concentration of property and income in the agricultural, industrial, banking, and service sectors intensified; profits from industry were not reinvested with the result that industries failed to expand and productivity fell. Most important, this model was inherently pro-urban, and it severely undermined the

---

[2] The program ran into budgetary and other restrictions, and most of the people involved in setting it up had left the ministry by 1985. It still continues, largely in conjunction with the National Population Council (*Consejo Nacional de Población*—CONAPO), which, in the Mexican government, is the main policy-making body in charge of women's programs.

peasant sector. Government investment in agriculture declined, and the few investments made were directed toward the establishment of irrigation systems for large-
scale capitalist agriculture. The terms of trade also worked against the agricultural
sector, and farm incomes lagged well behind urban incomes. Yet another way the
peasant communities subsidized the expansion of the Mexican urban economy was
through migration of young women and men to the cities. Industrial growth slowed
down, and "by the end of the 1960s, income distribution became increasingly
skewed, the current account of the balance of payments went into deficit, and public finance became unbalanced. Growth with price stability turned into its opposite,
stagnation with inflation, in the 1970s" (Aranda/Blanco 1981: 297).

The 'Mexican Miracle' was over. The rural crisis was first evident in agricultural production where the annual growth rate fell from 4.2 % in 1955–1965 to
1.2 % between 1967 and 1970 and 0.2 % during the 1970–1974 period. Thus the
average annual growth rate between 1965 and 1974 was less than 0.8 %, and it
continued to decline in subsequent years (Luiselli/Mariscal 1981: 440).

> During the second half of the 1970s, the crisis in agricultural production which began in
> 1965 and, fueled by low relative prices continues today, hit the national economy in two
> particularly sensitive areas: it increased inflationary pressures, brought on by the unprec-
> edented rise in food prices and raw materials from the rural sector, where supply could not
> satisfy national demand and, as the economy became food importing, as opposed to food
> exporting, the already serious deficit in the trade balance –i.e. of foreign exchange needed
> for economic growth– worsened (Luiselli/Mariscal 1981: 440).

These were only the immediately observable effects of the crisis; it also brought
about a shift in cultivation patterns. Mexico was no longer self-sufficient in food
production—in growing the corn, beans, and wheat that are the dietary staples of
low-income groups and are therefore socially important. An indicator of the agricultural crisis is that between 1970 and 1975 the area cultivated in corn decreased
by more than two million hectares (Montes 1981: 59). This shift restructured the
sector in favor of export-oriented agriculture such as beef production and horticul-
ture at the expense of basic grains (Rodríguez 1980: 65).

The cause of the agricultural crisis that began in 1965 can be found in this
restructuring of rural production, which is directly related to the position of the
Mexican agricultural sector in the international division of labor. In other words,
changes in the world economy affected the Mexican economy as a whole and the
agricultural sector in particular, reorienting it to the production of cash crops for
export. This shift has had important economic and social effects on small farmers
and, therefore, on rural women.

8.7 Rural Women and the Agricultural Crisis

The agrarian transformation of recent decades, in which capitalist agriculture displaced the traditional food crops cultivated in the peasant economy, has also changed
the way low-income rural women participate in production and social reproduction.

Recent history has shown that highly centralized industrialization policies—which create a surplus labor force in peasant economies—, first affect women. The rural exodus in much of Latin America, and especially in Mexico, is primarily female: women from rural communities migrate to the urban areas in search of work in the informal and service sectors. Rural women have thus played a central role in shaping the three basic characteristics of recent Mexican development: the rural exodus, the expansion of the urban service sector, and the growth of the informal sector of the economy. The imbalances created in rural areas by capitalist development have altered the social reproduction of the peasant economy and, more recently, have led to the expansion of a rural female proletariat. However, the heterogeneity of women's roles—the fact that they are members of peasant families, workers, and women—has hindered analysis of the differential effects of this process on them.

As members of peasant families, women must contend with the sharp decline in farm incomes that result from low world prices for agricultural products and from national policies that drain the surplus from the rural sector to finance urban industrialization. This situation is made more acute by the present financial crisis; the peasantry is expected to help repay the foreign debt by increasing production of export crops, which means that families must buy their food in the open market. For women, this expectation has meant heavier workloads while family health and nutrition deteriorate. Women have also had to compensate for this unequal exchange, either by intensifying their unpaid labor on the family plot, entering the wage labor force, or reducing their own personal food consumption.

As wage workers in agriculture or agro-industry, young rural women face unfavorable conditions. Because they lack legal and union protection, are discriminated against in the labor market, and are conditioned to behave in a docile manner, young women are more vulnerable than men to exploitation. Moreover, they face a fluctuating labor market controlled by middlemen and agents who often demand sexual favors in exchange for work.

As women, peasant women are still responsible for feeding, socializing, and protecting their children and families under very precarious economic conditions and often without the support of their husbands or partners, who have migrated. Women, moreover, are exposed to sexual violence both within the home and outside. Their gender subordination in social and political life makes it even more difficult for peasant women to improve their situation.

These problems are all interrelated: they do not affect women as isolated individuals nor are they derived from subjective issues. The fundamental problem is the larger process of subordination and exploitation of the peasantry, which is superimposed on women's gender subordination. Thus, the situation of rural women cannot be analyzed exclusively in terms of gender—the fact that they are women—nor can they be treated only in terms of their socioeconomic role—the fact that they are peasants.

8.8 Agrarian Structures, Women's Work, and Capitalist Development

The position of rural women in Mexico cannot be understood without taking into account the heterogeneity of their situation. Failure to do so can lead to abstract stereotyping of 'peasant women' and to programs whose superficial priorities do nothing to mitigate the fundamental problems brought about by changes in agricultural production patterns.

In the division of labor by gender, peasant women are in charge of all the unpaid labor of reproduction, which in agrarian societies is much heavier and more strenuous than that performed by urban women. Also, although all rural women participate either directly or indirectly in agricultural production, how and to what degree they do so is determined by the type of productive unit to which they belong. In Mexico, there are currently three types of productive units, each of which corresponds to a stage of the process of agrarian capitalist development.

For analytical purposes, the three types of family production units—subsistence, semiproletarianized and proletarianized households—can be differentiated according to the extent to which each is dependent upon the market to satisfy its needs. The gender division of labor provides the focal point for the following description of the kind of changes these units have undergone.

Since the 1930s, land distribution under the Agrarian Reform has provided for the reconstitution of subsistence peasant economies in the majority of the Indian communities and *ejidos*. The gender division of labor tends to be rigid, based on traditional cultural norms, and the role of women is oriented primarily toward the production of use values for family consumption. Women are exclusively responsible for the work of reproduction—all the activities that contribute to the reproduction and maintenance of the family labor force. These activities include storing, preserving, processing, and preparing food; socializing and educating children, providing psychological and medical care; and performing domestic chores. Women are also responsible for reproducing social networks—a very important activity that maintains solidarity and communication in subsistence communities. These activities range from visits and exchanges with family and extended kin in urban neighborhoods to performance of community ceremonies and collective rites.

Whether women participate in farming activities depends on the internal composition of the family labor force. Some tasks, such as carrying food to the men working in the fields, planting and harvesting (especially corn), and caring for the smaller livestock (chicken and pigs), are considered exclusively women's work. Women's participation in other activities, such as plowing, weeding, irrigation, and transportation, depends on the availability of male family members or community labor exchanges with relatives. The extent to which women are responsible for feeding and caring for cows or horses also depends on whether there are male relatives to perform these tasks.

Hewitt de Alcántara (1979) has suggested that the position of women in Indian communities tends to be better than that of the *mestizo* women in peasant communities. This suggestion does not imply, however, that women and men have equal positions in the subsistence peasant economy. The gender division of labor throughout rural communities is asymmetrical.

This division means that when there is extra work in areas considered 'men's work'—when productivity has to be increased to compete with capitalist farms or when a husband or son has migrated to another area—women are naturally expected to fill the gap. But when there is an overload of 'women's work'—when children are at school or daughters migrate and there is no one at home to help with the domestic chores—men only rarely try to help with household chores. As the economic crisis deepens, an invisible overload of work for rural women develops that demands that they expend greater physical effort and work longer hours than men.

A very important theoretical point emerges from this situation. The proposition that capitalist development, by subordinating workers to the interests of capital, is responsible for women's subordination is untenable. Gender subordination exists in traditional peasant economies even though women in indigenous cultures tend to have a higher social standing than their counterparts in *mestizo* society.

8.9 Women and Demographic Growth

In the 1930s and 1940s, the success of the agrarian reform was reflected in higher farm incomes and better levels of health, nutrition, and education among peasant families. Mortality, especially infant mortality, decreased significantly. Because fertility did not decline at an equivalent rate, population growth rates increased. New lands were opened for cultivation, and farming was intensified on land already in production. Although larger family size meant a heavier domestic workload, the more children in a family the more help women had with household and farm chores. In addition, higher incomes enabled families and peasant communities to support widows, divorcees, and single mothers. Since labor was needed to produce food, to build cities, and to create industries, all children and hands were welcome.

The tide began to turn in the 1950s, and the situation worsened in the 1960s when capitalist development intensified in the countryside. As the peasant economy became monetized, rural families had to purchase the labor, services, and goods that they had previously obtained through reciprocal exchange (Young 1978). As relations of reciprocity broke down, some families began to acquire economic advantages over others. More advantages accrued to those with more children as long as agricultural expansion and demand for cottage industry products and handicrafts—almost always the responsibility of women—were sustained. This fact clearly acted as an incentive to large family size, so that despite the decrease in mortality rates, peasant families did not take steps to reduce their high fertility rates.

When the public investments supporting *ejido* and Indian community farming began to diminish and family incomes fell, young women were sent out to work in the cities so they could send back remittances. Migration thus became a strategy for peasant families to obtain cash income as the rural economy became increasingly monetized and the demand for rural labor remained stagnant.

Young women were the first to migrate. Discrimination against women in agricultural, commercial, and service activities in rural areas meant that male children had better local employment opportunities than female children. Furthermore, the growth of the cities, and especially of the urban middle and upper classes, created a large demand for domestic servants—a secure and stable occupation for female migrants to the city.

8.10 The Proletarianization of Peasant Smallholders

Migration of some family members provided a survival strategy for core peasant families because it allowed them to subsist in the countryside, even as living conditions continued to worsen. At first, the strategy served to complement household farm and handicraft income, and daughters were the main members to migrate. By the beginning of the 1960s, however, migration became a sine qua non for rural social reproduction, and sons also began to migrate but most frequently to other rural areas where agricultural work was available or to the United States. A pattern of relay migration was thus established (Arizpe 1985). Because mothers and older women were left with a minimum of assistance, either female or male, their workload increased to intolerable levels: women were now responsible for agricultural and livestock chores in addition to the work of reproduction.

One of the main problems confronting women small farmers is getting the state to provide them with the credit and technical assistance to which they are legally entitled. Often a woman's brother, cousin, or nephew uses this as a pretext to take control of her allotted land. In any event, deprived of access to credit, technology, and other inputs, women are prevented from substantially improving the productivity of their land.

Older women who have no access to land as protection are often driven out by their families because of their low household incomes. This situation has led to a new and most cruel form of rural female migration—that of women over the age of fifty.

Semiproletarianized peasant households show a very clear pattern of female migration by age (Arizpe 1978). After the stage of greatest female migration—of single women between the ages of fifteen and twenty-two—there is a slowdown in the rate of female migration, particularly of married women up to the age of fifty. After this age, migration of both married women with their husbands and widows increases slightly. These older women end up either begging for a living or 'imposing upon' a relative in the city.

As a result of this pattern, there is an important intergenerational differentiation in women's work. The workload of mothers in reproductive and productive activities tends to increase over their life-cycle because their daughters are now integrated into the labor market. Any agrarian policy or development program directed toward women in semiproletarian households—now the majority in the Mexican countryside—must take these generational differences into account.

This semiproletarianization process of rural families fosters greater heterogeneity in women's productive participation, ranging from increased unremunerated family labor on the land to full integration into the wage labor market. This observation concurs with the findings for the Andean region where León and Deere (1982) have shown that no unilinear or determinant relationship exists between the process of capitalist development and the gender division of labor in production. In their study León and Deere (1982; see also Deere/León/Rey 1982) showed that women's participation in production varies according to the specific tasks that need to be performed, forms of labor recruitment, and the class position of the family. Further studies of semiproletarianized households in Mexico are needed to specify the changes in women's participation.

8.11 The Proletarianization of Rural Women

The single most important process affecting rural women in Mexico since the 1960s has been their entry into the rural labor market. The percentage of women in the rural wage labor force rose from 2.8 % in 1970 to 5.6 % in 1975. Although most women wage workers belong to peasant households, the pattern is changing, and many now come from landless proletarian households.

Mexican women become proletarianized in one of four ways. In one form, the household remains in the community, and all family members seek wage work in the immediate area. Mothers and daughters work either in agriculture or in sporadic jobs almost always related to domestic service. The second form also involves the family members' remaining in the community and doing wage labor at home in a variant of the putting-out system. This type of wage labor, which involves mostly sewing and assembly of consumer goods, is shifting from urban to rural households. Work is subcontracted to rural women by middlemen who come from the cities. This new kind of home work provides an alternative to migration.

The third option involves seasonal migration toward the regions of commercial agriculture. An important feature of this migration is that the women usually travel with their families. These people, whether married or single, live under the worst conditions, receive the lowest wages, and suffer from constant physical exhaustion, poor food, and broken family lives (Díaz/Muñoz 1978; Roldán 1982). Their children cannot attend school, and there is a high incidence of alcoholism and desertion by spouses. No legislation has been passed to protect these agricultural laborers; they are not unionized, and no government programs address either their needs or those of their children. Because they are often subjected to abuse and sexual harassment, young women prefer not to work as migrant agricultural workers.

Migrant women employed in agribusiness—located mostly in the northwest zone of the country—work under similar conditions (Roldán 1982). Young peasant women employed in agro-industry in their own region work under slightly better conditions. They live in their own homes (rather than in barracks), and although their wages are low and they receive practically no benefits, at least going out to work opens up a whole new world for them (Arizpe/Aranda 1981). This research has shown, however, that even when women are wage earners in their own right, they are not necessarily guaranteed more autonomy or authority within the family.

The massive entry of rural women into wage labor makes it imperative that legislation be enacted that takes into account their economic and social conditions and that programs be set up to provide them with technical and educational training. Of particular concern is the fact that the majority of jobs now available to rural women, including home-based work, are quite precarious. Employers take advantage of the fact that women get married or become pregnant to dismiss them; often as a result young women who have already left their families or communities become drifters with little hope of finding permanent work. One consequence, common all along the US border, is that the women end up in prostitution.

These trends suggest, especially as the financial crisis deepens, that the proletarianization of rural women and their families will continue to increase. The priority given to export agriculture, together with greater control by multinationals, will drive more women into the wage labor force. What is alarming is that the proletarianization of rural women still remains 'invisible' both to the state, responsible for agrarian policy, and to the social institutions in a position to offer rural women assistance.

8.12 Conclusions

Some important conclusions emerge from the foregoing analysis of the impact of agrarian policies on the situation of rural women. First, a given policy will affect rural women differently, depending on whether it is conceived of as an 'agrarian' or an 'agricultural' policy. As explained in the first part of the chapter, the Mexican agrarian reform established the *ejido* as an integrated social organization: by redistributing land, it strengthened the peasant economy as a way of life, not just as a specific productive activity. Although women were not designated as beneficiaries of specific programs—because the policy supported the family and the community as basic social units—the gender division of labor was subject to the equilibrium inherent in the peasant economy.

Thus, although women were still exclusively responsible for the tasks of social reproduction, they had some advantages in terms of social position. For example, they engaged in income-generating activities such as cottage industries, handicrafts, and petty trade. The labor exchange network—a mechanism that maintained labor force equilibrium among households in the community—enabled them to avoid an overload of work. Likewise, the community sheltered and protected widows, divorcees, and single mothers.

The process of proletarianization ushered in a new era in the gender division of labor, which became markedly differentiated. The incursion of manufactured goods and large-scale trade into rural communities deprived women of many of their income-generating, home-based activities, and, with the reduction in peasant income levels, the need to obtain cash income became a priority, precipitating the migration of young women to the cities. The workloads of senior female household members increased, and young women faced a discriminatory and exploitative rural labor market.

Given this situation, an agricultural policy focusing exclusively on productive activity not only does not support women but actually makes them economically and socially more vulnerable. An agricultural policy that gives priority to production of export crops might create female employment, but by being defined narrowly as an 'agricultural' policy it does not take responsibility for the deplorable working conditions prevalent in the majority of the jobs it generates. Its aim, as a policy, is only to sustain economic growth or to service the debt of an urban and industrially oriented development model. By the terms of this model, the fact that rural women have suffered rapid proletarianization is only incidental to agricultural policy.

The aim of an agrarian policy based on land distribution is, on the other hand, to develop peasant society and enable it to supply the rest of the country with the necessary food and raw materials. This type of policy must provide rural women with the resources they need to be subjects of rural development and agents of their own social transformation rather than objects—disposable labor—of a growing economy benefiting only urban elites.

An agrarian policy, therefore, must support the productive activities of rural families and generate new productive activities for young women while providing them with the necessary technical and professional training. *Ejidatarias* and female smallholders must, in turn, be assured of access to credit, inputs, technology, and extension services.

In sum, to the extent that a peasant economy is still viable, the impact of agrarian policies upon rural women will largely depend upon the internal equilibrium of households and communities. However, since rapid proletarianization is well under way, a policy that fails to include programs specifically targeting women will only put them at a further disadvantage in the rural labor market.

When these policies are narrowly conceived as agricultural rather than agrarian, their negative impact is increased. The result of ignoring the serious social consequences of promoting commercial—and above all export—agriculture is the illegal and deplorable conditions under which proletarianized rural women currently must work.

Until there is an agrarian conception of rural development that explicitly includes rural women and until the labor legislation and technical and social support they need is forthcoming, the peasant social base in the countryside will continue to erode. As long as this erosion continues, the future confronting rural women will be one of increasing exploitation, alienation, and economic insecurity.

References

Aranda, Ximena; Blanco, José, 1981: "El Desarrollo de la crisis 1970–1976", in: Cordera, Rolando (Ed.): *Desarrollo y Crisis de la Economía Mexicana* (Mexico City: Fondo de Cultura Económica): 297–335.

Arizpe, Lourdes, 1978: "Mujeres migrantes y economía campesina: Análisis de una cohorte migratoria a la ciudad de México 1940–70", in: *América Indígena*: 38,2 (April–June): 303–325.

Arizpe, Lourdes, 1982: "Relay Migration and the Survival of the Peasant Household", in Balán, Jorge (Ed.): *Why People Move: comparative perspectives on the dynamics of internal migration* (Paris: UNESCO): 187–210.

Arizpe, Lourdes; Aranda, Josefina, 1981: "The Comparative Advantages of Women's Disadvantages: Women Workers in the Strawberry Export Agribusiness in Mexico", in: *Signs*, 7,2 (Winter): 453–474.

Barbieri, Teresita, 1981: "Un estudio de dos UAIM" ILO, Mexico, mimeo.

Chávez Padrón Martha, 1974: *El Derecho Agrario en México* (Mexico City: Porrúa).

Deere, Carmen Diana; León, Magdalena; Rey, Nohra (Eds.), 1982: *Debate Sobre la Mujer en América Latina y el Caribe: Discusión acerca de la Unidad Producción-Reproducción* (Bogota: ACEP).

Díaz, Rönner Lucila; Muñoz, María Elena, 1978: "La mujer asalariada en el sector agrícola", in: *América Indígena*, 38,2 (April–June): 327–334.

Fabila, Manuel, 1982: *Cinco Siglos de la Legislación Agraria* (Mexico City: Centro de Estudios Históricos del Agrarismo en México).

FAO (Food and Agricultural Organisation), 1983: "Examen de las Políticas y Estrategias de Reforma Agraria en México" (Rome: FAO).

Gutelman, Michel, 1974: *Capitalismo y Reforma Agraria* (Mexico City: Editorial ERA).

Hewitt de Alcántara, Cynthia, 1979: *La Modernización de la Agricultura Mexicana* (Mexico City: Siglo XXI).

León, Magdalena; Deere, Carmen Diana, 1982: "La proletarización y el trabajo agrícola en la economía parcelaria", in: Deere, Carmen Diana; León, Magdalena; Rey, Nohra (Eds.): *Debate sobre la mujer en América Latina y el Caribe: discusión acerca de la unidad producción-reproducción* (Bogota: ACEP).

Luiselli, Cassio; Mariscal, Jaime, 1981: "La crisis agrícola a partir de 1965", in: Cordera, Rolando (Ed.): *Desarrollo y Crisis de la Economía Mexicana* (Mexico City: Fondo de Cultura Económica): 439–455.

Macías Coss, Ruth; Zaragoza, José Luis, 1982: *El desarrollo agrario en México y su marco jurídico* (Mexico City: Centro Nacional de Investigaciones Agrarias).

Montes de Oca Luján, Rosa Elena, 1977: "La cuestión agraria y el movimiento campesino 1970–1976", in: *Cuadernos Políticos*, 14 (October–December): 57–71.

Paré, Luisa, 1988: *El Proletariado Agrícola en México:¿Campesinos sin tierra o proletarios Agrícolas* (Mexico City: Siglo XXI).

Reyes Osorio, Sergio, 1972: *Estructura Agraria y Desarrollo Agrícola en México* (Mexico City: Fondo de Cultura Económica).

Rodríguez, Gonzalo, 1980: "Tendencia de la producción agropecuaria en las dos últimas décadas", in: *Economía Mexicana*, CIDE, 2.

Roldán, Martha, 1982: "Subordinación Genérica y Proletarización Rural: Un estudio de caso en el Noroeste Mexicano", in: León, Magdalena (Ed.): *Las trabajadoras del Agro* (Bogota: ACEP): 75–101.

SRA (Secretaría de la Reforma Agraria), 1985: *Ley Federal de Reforma Agraria* (Mexico City: SRA).

Young, Kate, 1978: "Modes of Appropriation and the Sexual Division of Labor: A Case from Oaxaca, Mexico", in: Kuhn, Annette; Wolpe, Ann Marie (Eds.), *Feminism and Materialism* (London: Routledge and Kegan Paul): 124–154.

ANTROPOLOGÍA BREVE
DE MÉXICO

Coordinadora: Lourdes Arizpe

Autores: Arturo Argueta, Lourdes Arizpe, José del Val, Guillermo de la Peña, Joaquín García Bárcena, Mercedes González de la Rocha, Claudio Lomnitz, Linda Manzanilla, Eduardo Matos, Noemí Quezada, Carlos Serrano, Héctor Tejera, Patricia Torres, Leopoldo Valiñas

ACADEMIA DE LA INVESTIGACIÓN CIENTÍFICA

CRIM

Book cover of *Anthropology of Mexico: an Overview*, 1993, edited by Lourdes Arizpe, with texts of 21 Mexican anthropologists, presented at the International Congress of the International Union of Anthropological and Ethnological Sciences which she organized in Mexico City in 1993. The Photo was taken at CRIM-UNAM which granted permission.

Part III
Social Development

Chapter 9
A Society in Movement: Anthropology of Mexican Development

9.1 Introduction

In the 'long count' of millennia, as the Mayans counted, the people who inhabit the Mexican territory were a society always on the move.[1] In the dawn of the settling of the Americas, small bands of migrants who had trekked over from Africa to Asia crossed the Bering Strait and forged indigenous civilizations in lands that were much later to be called the Americas. In the 'short count' of centuries, migrants flooded in from Europe to seek their fortunes or to plant the seeds of utopias. And in today's 'count of days', Mexicans have taken their own culture and fortunes into their own hands to again transform their lives in the third millennium.

As if blown by a gale, geographical mobility began again in Mexico in the 1950s. Between 1950 and 1990, more than eleven million Mexicans left their ancestral villages and moved to other places. Since the sixties, commotion was added to the culture—and to politics—, so that today, at the end of the century, Mexico's cultural pluralism has opened up to global change. Thus, it may be said that Mexicans continue to be governed by the original sign of the last Meso-American era, the sign of *Nahui Ollin* or 'Fourth Movement' in *Náhuatl* (Aztec) language.

9.2 *Nahui Ollin*

If Meso-America/Mexico were the centre of the world, Europe would be the East and Asia the West. Since the planet is spherical, is it not then perhaps presumptuous to think of the world in terms of East and West? Or, in terms of civilization and

---

[1] Originally published in Spanish as the chapter "Una sociedad en movimiento" in Arizpe, Lourdes (Coord.): *Antropología Breve de México* (Mexico City: Academia de la Investigación Científica–CRIM-UNAM): 373–397. Unpublished in English.

L. Arizpe, *Migration, Women and Social Development*,
SpringerBriefs on Pioneers in Science and Practice 11,
DOI: 10.1007/978-3-319-06572-4_9, © The Author(s) 2014

barbarism, of 'West' and the 'rest'? Little by little, history has been undermining all kinds of 'centrisms'. 'Man' is neither the centre of 'Creation' (anthropomorphism) nor is the 'Earth' the centre of the solar system (geocentrism); nor is the 'West' the centre of the world (Eurocentrism); nor is 'man' the centre of society (androcentrism); nor is *Homo sapiens sapiens* the centre of the natural world (anthropocentrism). How can we think of ourselves in a seriously de-centred world?

The autochthonous settlers of this continent that today is called America had no generic name when they began their colonization of these lands, between the 26th and 12th millenniums before our era, as García-Bárcena (1993) has pointed out. Every semi-nomadic band, every ethnic group, frequently referred to themselves as 'the true men'—like the *halach unic* in Mayan. Exactly in the same way as the Romans did, for example, they mostly called their neighbours 'barbarians'—like the *Chichimecas* in *Náhuatl*. Apparently, this division between 'us' and 'them' is one of the cognitive principles that have organized life among peoples throughout all of history. East and West, North and South, directions almost always based on the cardinal points of the compass, yet, in several Meso-American cosmologies, there were two more directions: up and down.

Thus, we human beings tend to construct identities superimposing a grid of 'us' and 'them' at every step of the way. Let us think, for example, of a mythical identity of us, the women, and them, the men; but immediately after that, of us, men and women of the neighbourhood of La Conchita (an Aztec *'calpulli'* presently called *'barrio'*) and them, those of the *barrio* of Los Alacranes (another Aztec *'calpulli'*) of the village; then of us, those of La Conchita and Los Alacranes in the town of Coachochitlan and them, those from the town of Juchipatlan; then of us, those from these two towns in the state of Morelos (descendants of the Tlahuicas) and them, the *'Chilangos'* (descendants of the Mexica of Tenochtitlan) from Mexico City; then of us, the Tlahuicas and *Chilangos* as Mexicans, and them, the *'Gringos'* (US Americans); then of us, Mexicans and *'Gringos'*, the North Americans, and them, the Europeans. Everyone chooses her/his own level of identity.

One by one, each identity fits into the other, like fractal structures, just as the new fractal geometry teaches us to see the world in a new way. We had never been able to find the exact fit of local, national, regional or continental identities, because, on the basis of a Euclidian geometric model, triangles and cones and spheres simply cannot fit into each other. On the other hand, every fractal contour contains within itself many of its own forms at different scales.

In short, identity has a fractal structure in social space, and a history in time. In other words, present-day science teaches us that *every concept depends on the place in space and the time frame in which it is situated.* That is perhaps the most important message offered by anthropology at the end of the millennium. Who is the person that is gazing and from what vantage point is she/he gazing? How does this person interpret and translate the culture under observation? All of which leaves us, as Geertz (1973) would say, like the natives of Polynesia, navigating on the high seas, having discarded old nautical routes and consoling beliefs, to face with our new technologies an unknown civilizational destiny.

9.3 Migratory Cultures

Settled in the cone of abundance formed by the Sierra Madre in Mexican territory, the prehistoric settlers interacted with the wealth of biotic and geomorphologic resources and created a very high linguistic and cultural density. Crafting
and manufacturing, urban life and continuous long-distance exchanges began to
appear, as described by Manzanilla (1993). Empires were consolidated (Matos
1993), and the cities were wrought with splendour. This was how Meso-America
was born.

Those first migratory settlers did not, of course, cease their wanderings in
and through Meso-America, as many of them continued their migration into
Central America and beyond, until they reached the end point of that migration in Tierra del Fuego. On the way, when confronted by the challenge of the
pyramidal ecology of the Andes, they created the high cultures that culminated in
the Tawantinsuyo, among others, and when confronted by the maelstrom of the
Amazon jungle, the agile riparian communities of those forests. However, filaments of continuity with all of these people exist in genes and phenotype in the
Americas, as has been pointed out by Serrano (1993).

In spite of these continuities the borders of Meso-America, as established by
Kirchoff (1943), did create a cultural commonality: out of a great cultural diversity
foundational cultural elements were taken to form a distinct cultural current. By
the year 1,500, according to Valiñas (1993), it is estimated that there were approximately 120 languages spoken in the territories of Meso-America and Arido-
America. There was a greater linguistic diversity than exists at present.

Cultural patterns are closely linked to geographical mobility. In Meso-America,
after a long period of almost ten centuries of civilizational flourishing around
large ceremonial cities encircled by many agrarian villages, conflicts arose, some
of them internal, perhaps quarrels over food surpluses and over access to natural
resources, and others external, having to do with conflicts involving plundering
and the invasion of antagonistic ethnic groups. These conflicts brought about the
downfall of these empires and their social and political organization ruled by theocratic elites.

And the farmers and artisans, like waves after a violent flood that subsides in
foaming fringes, returned to the tranquillity of their scattered agrarian communities. For several centuries during the colonial era and after independence, there
was a periodic pendular movement between attraction towards emerging cities and
a reflux towards rural villages after the decline of the cities, an ebb and flow and
reflux seen throughout the history of Meso-America and Mexico.

This mobility between the countryside and the cities always crossed over the
invisible, but nevertheless effective, boundaries that, since the colonial era, had
divided the republic of the 'Spaniards' and *criollos* from the republics of the
Indians. It should be made clear that this political boundary that was thought of
as being racial was also ambiguous. For example, the Tlaxcaltecas, having allied
themselves to the Spaniards, still say today that they were never 'conquered' and,

therefore, that that boundary never applied to them. In fact, they did enjoy special treatment by the Spaniards and therefore, during colonial times, they were a liminal group, that is, a group on the margins, in the history of Mexico. Liminality takes on its own characteristics as a condition of human groups.

That boundary between the indigenous peoples and the Europeans deepened during the first centuries of the colonial era due to the terrible mortality that weapons and especially epidemics caused among the Indians and which, because of its numbers, may be one of the largest genocides in the history of the world. There is no agreement among researchers on whether there were eleven million or six million Indian inhabitants when the Spaniards arrived, but it is known that the population had decreased to only one million by the year 1605, when one of the last epidemics occurred. Compare this figure with the estimates of the Jewish Holocaust (six million)—carried out deliberately—and with the genocide in slave trafficking from Africa (ten million).

Thus, there was relatively less geographic mobility and no social mobility in colonial times. The social pyramid, in order of hierarchy of '*peninsulares*' (Spaniards born in Spain), '*criollos*' (descendants of Spaniards born in New Spain), '*mestizos*' (mixed-race people), and Indians, was influenced territorially and geographically. What is the smallest link, the one that forms the first alliance that can transform the whole social structure of a society? Love, of course. That was the reason for the rigid rules about amorous relationships between men and women during the colonial era described by Quezada (1993). The Spanish colonialists tried to regulate and control what was inevitable, that is, relationships between women and men from different castes and ethnicities. Was it not perhaps those relations of love, whether secretive or scandalous, ephemeral or permanent, in constant conflict with the rulings of the colonial government and with peninsular and *criollo* racism, that little by little gave birth to a new *mestizo* Mexican society?

With independence, political domination by the *peninsulares* was replaced by that of the *criollos*, leading to the emergence of a disorderly struggle on the part of the *mestizos* throughout the nineteenth century to break down that untouched pyramidal structure of race, culture, and urban predominance.

9.4   The Revolutionary Movement

That structure of domination was finally destroyed when a new field of political negotiation was consolidated after the armed conflicts of the Mexican Revolution (1910–1917). After several centuries of being rooted in one place, the peasants, ranchers, ranch hands, agricultural wage labourers, industrial workers, the urban middle class, the '*soldaderas*' (the women who followed the revolutionaries) and the people in the haciendas all left their lands to go and fight. It was a sudden movement, the '*bola*'—the 'rolling ball' as they called it—carried them away

Table 9.1 Rates of growth between censuses, 1910–1980

Year	1921	1930	1940	1950	1960	1970	1980
Rate	−0.50	1.78	1.77	2.69	3.08	3.40	3.21

Source CONAPO (1988: 62)

and created a new country. It broke down racial barriers, preserved independence and national territorial integrity, recognized the Indian past and created a *mestizo* future, and succeeded in agglutinating a mosaic of regional, cultural and ethnic groups, immigrants and people from the emerging classes under the banners of nationality, social justice, education, and democracy. Once it had become consolidated—and negotiated in silence with the regional chieftains, the United States, the Church and the new regime—the throngs of people returned to their homes, back again to the countryside.

The exception to this was the new political class, which came from different parts of the country, especially from the north, settled in Mexico City, and … repeated history. Anáhuac-Tenochitlán moved across the mirror of time to re-emerge in the Valley of Mexico Federal District with all the ambivalence of a centralism that was to concentrate financial and human resources in that city in the next six decades, until it became the largest city in the world.

On their part, the revolutionaries who followed Emiliano Zapata, both *mestizos* and indigenous people, returned to their corn fields, just as Tejera Gaona (1993) describes. After the Agrarian Reform which gave them back their lands, they actually reinstated the corporate community-based system of economic and political life that had been affected by the concentration of lands in the hands of the haciendas and the Church until the previous century. The distribution of lands and the opening up of new lands to cultivation fostered an *agricultural miracle*, creating cheap staple food and financial surpluses that were then used to develop industry and urban centres, especially Mexico City. The improved production and dietary conditions and the new health campaigns that accompanied rural education also fostered a *population miracle* that brought about a decrease in mortality, especially infant mortality, which dropped from 132.0 in 1950 to 53.1 in 1980 (CONAPO 1988: 62). With a decreased mortality rate but with the same level of fertility, the country's population grew at a very fast rate, as can be seen in Table 9.1.

With more children surviving, new processes began to occur in agrarian communities. On the one hand, in the regions in which Agrarian Reform had given out plots of at least five hectares, they began to be subdivided in each generation, leading to very small plots of land for cultivation. In some regions this pressure on the land led to the continuation of demands for the redistribution of large landholdings whose ownership was concealed through a variety of means, and which gave rise to the 'battle of the countryside' in the sixties and seventies. But in many regions in central and southern Mexico there were no longer good lands to be distributed. In addition, in many others, such as in the Mixteca, Chinantla and Mazahua zones, deforestation had already brought about a constant, unrelenting erosion of lands that made it no longer possible to cultivate them. From many of these regions, a massive rural exodus began in the fifties and continued during the following decades.

At the same time, cultural change accelerated. After the Mexican Revolution, rural schools, cultural missions and increasing extensions of the educational system brought literacy and schooling to rural communities. But education has always been biased towards urban life; in fact, it exalts urban life and denigrates rural life and everything having to do with the life of the peasantry and indigenous peoples. So, was it not to be expected that such schooling would add another push factor for escaping from rural communities? And was this not expected to happen when economic and cultural success began to depend on having an educational level to provide access to better-paid jobs? Specifically, Indian communities were the ones most affected by a policy of cultural integration that encouraged cultural assimilation. Both factors, therefore, fostered rural outmigration.

9.5 Rural–Urban Migration

In the forties, as the demand for manufactured goods rose as the United States became involved in the Second World War, Mexico expanded its industrial plant. The new jobs created in Mexico City, where investment capital, factories and markets began to be concentrated, became the pull factor that attracted young people. They began to arrive, as De la Peña (1993) describes, first from the medium-sized cities and then from rural communities. As early as 1950, 12.9 % of the Mexican population lived in a different state from the one in which they had been born. Most of the migrants, 1.14 million, came from the states around Mexico City, although 1.021 million came from the states in central-western Mexico, particularly from Michoacán and Jalisco.

It was the peasant and Indian communities that provided the young workers and cheap staple foods that launched Mexico's industrial take-off, among them the young women who had always been in the majority in migratory movements to the cities. These young people, brought up, fed, cared for and schooled in rural villages, whose role is described by Tejera (1993), were the ones who subsidized the formation of the labour force for Mexican industry in Mexico City and in other cities.

Since the mid-fifties, agricultural and fiscal policies also drew financial resources from the agrarian sector to foster urban growth, and the market made the prices of manufactured goods rise relatively more quickly than those of agricultural products—particularly corn (maize), whose rise in price was suspended from 1957 to 1973, while other prices were rising. As a result of such subsidies to cities agricultural livelihoods were destroyed.

Several strategies were used by farming families in an effort to survive. By the end of the sixties, migration was being fuelled by push factors and population growth. In this second wave of migration it was the poorest strata of rural families who began to migrate from states such as Tlaxcala, Oaxaca and Chiapas. Thus, beginning in the seventies, rural urban migration in Mexico became massively widespread among peasant and Indian peoples.

Faced by constant deficits in their budgets, small landholders began to follow a strategy of relay migration. First, the father migrated; then usually the oldest daughter or, in some cases, the eldest son; when they married, the second daughter or son were sent, and so on successively with all siblings. In this way, the family was able to always have one or two children migrants sending back cash remittances.

Interestingly in this economic movement, because of their cultural tradition, Indians and 'campesino' families followed corporate strategies in migrating. The first 'paisanos' (members of the community) who were able to settle in the city then brought the children of their families; also, a niece or a cousin or a foster-child or someone who was a close family friend would then also be sent to the city. In some villages, even a 'correo' was set up whereby a person of the village would travel early every Sunday from her/his community and stay all day at the bus station receiving the messages, remittances and even dirty clothes that s/he would take back to the families in the village. At the same time she or he would bring back all the news about how everyone was behaving in the city.

All of this movement began to create an urban life in Mexico that was very different from what it had been before. The urban labour market was transformed, absorbing the migrants, as described by De la Peña (1993), to create a broader popular and working-class culture with new ways of life, like those indicated by Torres (1993). Also, the social organization was transformed, as the study by González (1993) shows, with a new combination of nuclear and extended families and with new strategies for survival through a new economic and social participation by women.

9.6 Women in Movement

In the sixties a profound change began in women's participation in other countries as well as in Mexico where it had specific characteristics. Firstly, rural–urban migration gave many women freedom of geographic mobility; secondly, the contraceptive pill, and the view that it was no longer necessary to have so many children, gave them greater freedom from their biological role as mothers; thirdly, feminism gave explicit expression to their old and new desires to broaden the range of action of their lives towards remunerated work, educational attainment, personal realization, and political participation; fourthly, the expansion of the labour market and the educational system and labour markets in an urban industrial society have allowed them to open up for themselves unprecedented horizons of knowledge and personal economic autonomy.

Talking about cooking with Amuzgo women in Tlacoachistlahuaca, 2008. The photo was taken by Edith Pérez Flores and reproduced with her permission

The emancipation of women has been conceived, and rightly so, as the last great social movement of the twentieth century. Other movements await us in the new century, but there is no turning back for the women's movement, *the longest revolution*, as it has been called. There is, in addition, a new millennium in which to prosper.

Let us begin by explaining the change in the movement's roots, with peasant and Indian women. As the pillars of the family agricultural production unit, their direct and indirect participation in production has sustained the agricultural miracle in Mexico, and their tireless work in the family has made possible the forms of migration that have transformed the countryside and the city. In some regions, the migration of their husbands and children left them in charge of the small landholding whilst having to cope with discrimination in access to land titles, credit, technical assistance, and access to new agricultural technologies. It is well known that as long as the mother is still living in the house in the village, everyone can return to it at any time; however, when the mother migrates, that home irremediably disappears from the community.

It has been the daughters of these women, the generation that grew up in the sixties, who have woven the social networks that have maintained the cohesion of Mexican society throughout so many changes. They were the first to go into wage-earning work in commercial agriculture, by picking cotton, tomatoes, and strawberries. Many of them at first went out alone to wage labour, and then in the eighties with their husbands and children. Today there are a huge number of women agricultural labourers who in fact face the worst working and living conditions in the country. After that first stage as agricultural labourers, rural women went into wage-earning work in agribusiness, processing tomatoes and strawberries for shipment and export. In the third stage, from the seventies to the nineties, many of them migrated to the border with the United States, to work in the *maquiladoras*, the impermanent assembly plants for clothing and electric and electronic appliances. Many others stayed in their villages and today are doing subcontracted work, mainly weaving and sewing in their own homes.

As has already been said, women who migrated to the cities were domestic workers and a few were employed in manufacturing but they were also predominantly employed in the service economy and in precarious employment. Among the latter, those who stand out most are the Indian women, who prefer the relative freedom afforded to them by self-employment, especially in market vending and who are known as the 'Marías' in Mexico City.

In all of these jobs, the wages and working conditions vary and are mostly inadequate, but they have allowed thousands of women to develop and to acquire greater autonomy and a greater awareness of their possibilities in life. A major concern is that this new participation of women in remunerated work has created a *triple work day*, with the paid work added to their family and domestic responsibilities. Little by little, customs are changing so that men are sharing more of this work, both in caring for the children and the elderly and in housework. Thus, as has happened in many other countries, the traditional division of labour by gender is changing within the household.

The new awareness fostered by feminism has led to the increased participation of women in urban movements, in which they are leaders, in '*campesino*' and Indian movements, and in political parties. This new participation has generated complex debates about the prioritization of feminist claims and general political demands; and about different gender behaviour, that is, whether women tend to be less corrupt, are gentler in exercising power but more rigid in decision-making, more sensitive yet more impulsive in their actions.

In short, this new presence of women in academia, public administration, politics, mass media and art opens up fundamental questions about the nature of society. In the seventies, men and women were thought to be equal and thus, at that time, absolute equality between them was demanded. In the nineties, women and men are perceived as having different predispositions; given this, it is no longer a question of forcing women to fit into the political and cultural frameworks created by men, but rather of changing those basic frameworks of society to foster a balanced and more harmonious participation by both genders. If we put aside for a moment the complexity of these relations, we should remember that the final intention is unobjectionable: to raise the level of possible happiness for society as a whole.

9.7 Indian Movements

By the end of the sixties, when peasant and Indian migrants were increasingly moving to the cities, driven by the crisis in small landholdings in agriculture, job creation in the cities began to slow down. More and more, migrants had to stay in the informal sector as masons, market vendors and porters in the Merced market, or street sellers like the 'Marías'.

Urban inhabitants, who thought that Indians were to be found only in a remote past, were surprised to see Indian women in the streets of their city. Dressed in their colourful outfits, speaking their ancient languages, they came to remind the Mexican state that Mexico is a pluricultural country. This historical re-encounter also had to do with the revitalization of the Indian organizations that were demanding a new development policy towards their peoples.

In 1975, for the first time in the history of Mexico, the First National Congress of Indigenous Peoples was held in Pátzcuaro, Michoacán. That conference was the spark that detonated a new awareness about Indian peoples in Mexico and the beginning of a new debate on cultural pluralism and the right to safeguard indigenous languages and cultures.

The vocabulary of the debate changed notably. It has already been mentioned that the term 'Indian' created a single identity for culturally very diverse peoples. Another term was also contested. History books say that the Spaniards came to Meso-America … The Spaniards? Our contemporary friends from the Iberian Peninsula tell us that, at present, there is now *a Spanish state*, which encompasses several autonomous identities and nations, mainly the Catalans, Basques, and Castilians. Once again, we encounter the sociopolitical polarity between 'us' and 'them'. Columbus introduced this historical ambiguity with the terms 'the Indies' and 'the Indians' based on his navigational error. Interestingly, both he and Hernán Cortés came from ambiguous identity situations. Was Columbus a Genoan, a Spaniard or a Portuguese? Hardly a Spaniard, since the Catholic monarchs had only very recently tried to consolidate the alliance of Castile and Aragon. Hernán Cortés seems to have been born in Old Castile, but to soothe his Gallegan and Andalusian soldiers who threatened perpetual mutiny, he always asked 'Malintzin', his 'Indian' woman translator, to speak only of 'Spaniards'. "'We' the Spaniards and 'you' the …".

Indians? No attempt was made to see and understand the extraordinary cultural richness that the invaders had before them. Only the Catholic missionaries, notably Fray Bernardino de Sahagún, were able to safeguard some of this richness. It was undoubtedly a very complicated endeavour. Firstly, to distinguish the strange Meso-American phonemes: for example, *Uitzilopochtli*, which was simplified to '*Huichilobos*'; or, as Valiñas (1993) notes, to distinguish between a denominative such as *Zapoteco* and *Zacateco*; a generic term, such as *Otomí* and *Chichimeca*; or a group that interprets us, like *naarinuquia* or *tzotzil*. They were to be 'Indians', all of them. Thus was born the term *Indian* as a colonial category.

We must understand, then, the unbearable heaviness of these terms because of the political and emotional load they carry. In 1992, on the occasion of the fifth centenary of the first voyage of Christopher Columbus, all these terms were scrutinized and disparaged to the point of satiety. Let us see: what is this about the 'discovery' of 'America' by 'Spain'? To begin with, the continent was already here, as historian O'Gorman (1958) famously and decisively argued; then, because of an academic confusion, instead of calling it 'Colombia', they gave it the name of a person who never set foot on this continent, Amerigo Vespucci. So, if it had not been for *Doña* Isabel's intuition, 'Spain' would never have been consolidated and Columbus would not have set his sails towards the West.

It is even worse to state that the 'Spaniards' 'conquered' the 'Indians'. Let us not repeat the question of who the 'Spaniards' were; on the other hand, let us add that in their undertaking, which was daring and skilful without question, they had tens of thousands of native allies and that unfathomable myth of the 'return of Quetzalcoatl from the East' that infiltrated the minds of the Mexica emperor. To cut this discussion short, nothing better than the words I heard at the 1975 First Congress of Indian Peoples from an indigenous leader: "A woman can be conquered but only with love. The Spaniards did not conquer us. We were invaded".

This very important shift in perceptions and in cultural policy has been described by Del Val (1993). It culminated with the insertion of the right to preserve indigenous cultures and identities into Article 4 of the Mexican Constitution in 1992, 500 years later.

9.8 Migration Towards the Northern Border

In the seventies, the routes for migrants from rural communities began to diversify. Many started migrating to other cities, especially to Guadalajara and Monterrey. Already, since the thirties, there had been important flows of Mexican migrants to the United States, under the Bracero Program, and later as undocumented immigrants. Many stayed in the border zone, where the *maquiladora* industries began to offer job opportunities. In Baja California Norte, for example, the population increased from 550,000 inhabitants in 1950 to 1,177,900 in 1980 (Chávez Galindo 1987); 63.4 % of these had been born in other states of the country (Pick/Tellis/Butler/Pavgi 1990).

However, *maquiladora* industries, after a first period employing men, began to shift towards employing women, so female migration greatly increased to most cities along the border. Studies showed that the reason for this preference is that, according to employers, women have greater manual dexterity, are more docile, do not join labour unions and, when they marry or become pregnant, they usually stop working or are forced to resign. This allows employers to have a constantly renovated labour force with important savings on promotion, maternity leave, seniority, and retirement costs. This feminization of migration to the border zone is reflected in the fact that the index of masculinity for migrants in 1940 was 1.17, whereas it dropped to 1.00 in 1960 and 0.93 in 1980 (Chávez Galindo 1987: 37).

Migrants had come predominantly from the states of Jalisco, Michoacán, and Nayarit as well as Sinaloa, Zacatecas and Durango. By the eighties they increasingly came from the central and southern states. This new flow included Indians from the southern states, especially Mixtecs and Zapotecs from Oaxaca, who have established clear social routes between their communities and the destination sites for migrants in the United States. Indian groups, it must be said, who already lived along the border, such as the Pima and Pápagos, had historically crossed the border at all times.

Studies have estimated that, since the eighties, as much as 80 % of household income in the Mixtec region of Oaxaca has come from migrants living in cities in both Mexico and the United States. Once again, exchange networks have been created between the communities of origin and the places where the migrants work in the neighbouring country. This vast economic and social grid is rooted in millions of Mexicans who now live in the United States and is changing the perception of the country's geographical and cultural border.

To begin with, it has fostered the creation of new forms of cultural expression. On this side of the border, new youth cultures have emerged, such as the 'Cholos', groups of young people in the border zone who want to distinguish themselves by the way they dress, talk, and dance. Through their body and dress language they send the message that they 'belong' to a group, and are 'different' from other youth groups. Once again, we encounter the 'us' and 'them'. All groups of urban youths in cities do the same, except that the Cholos can select cultural signs from a double repertoire: Mexican and American, especially the American South-west cultural tradition. They often choose elements to show that they reject the inclusion/exclusion imposed on them by the political border and, on the contrary, they state that they are owners of their own social space, carved out by their own hands from the map of symbols that surrounds them.

9.9 The Chicano Movement

Today, there are at least three distinct groups of Mexicans in the United States: the first are the proud Mexican-Americans, whose genealogy may go back several centuries. They are descendants of families from New Spain that colonized California, Arizona, New Mexico, and Texas before the Americans arrived. In the first half of this century, many refugees from the armed conflicts of the 1910–1917 Revolution also migrated to these US states. They were later joined by the migrant workers of the Bracero Programme, who came mostly from Jalisco, Michoacán, and the neighbouring states.

These migrants took with them the rich and turbulent cultural effervescence of the post-revolutionary years and which created a boom in Mexican art (muralism), music and films, and the search for a cultural identity through Mexicanity, 'Aztequismo', and the culture of the Charro—the Mexican hacendado and cowboy. These cultural symbolic systems, transferred by the Mexican migrants to their American-born children, who were no longer agricultural labourers but now part of the American working class and even middle class, produced a distinct and original iconography of Chicano art; and the mixture of these cultural symbolic systems with the popular culture of the 'Anglos' led to a new combative language of their own that is expressed in fiction, films, and linguistic turns of phrase. This is the second group that arose from Mexican migration in the United States.

Finally, the third and fourth generation of Mexican migrants, many of whom have gone to university and now identify themselves as 'Hispanics' or Latinos in the United States. This group shares affective ties with their Mexican or Latin American ascendancy and the Anglo-American ideals of work, education, and democracy. Their desire to be close to the Mexican and Latino cultures is genuine but with an eye for differences; their incorporation into American political life is conscious and affirmative.

Through these processes, does Mexican culture threaten 'American national identity' as Huntington (2004, 2005) claims? Asking a question using the same simplistic view, will American culture end up destroying Mexican culture? Neither one nor the other.

We should remember that the tendency to talk in terms of 'our' culture is based on a perception that selects specific traits that we think identify our group compared with others; but there is always a continuum of cultural traits linking one group to another. The people who live in those 'borderlands' are the ones who hold in their hands the future definition of cultural trends in those liminal locations. Is Mexico itself not a nation of liminal cultures, that is, of cultures born in response to other cultures?

Lastly, in this last decade of the twentieth century, Mexico is open to a new migratory movement, this time along its southern border.

9.10 Migration Towards the Southern Border

In the south, the two Sierra Madre mountain chains become a knot, starting in the Mixtec region and from there moving down to Guatemala. This geography creates countless valleys, some with inaccessible basins and slopes that have preserved the greatest linguistic and cultural diversity of Meso-America down to our present day. Until recently, this region, incorporating the states of Oaxaca and Chiapas, seemed impenetrable given its geography—imposing mountains, endless canyons, and tangled jungles—and the long history of conflict in its inter-ethnic relations.

In spite of this history, today, the southern border of Mexico has witnessed great movement. Between 1970 and 1980 the indigenous population in the southern border region increased from 68,000 to almost 135,000 people. At present, the coffee plantations in the Soconusco no longer receive Tzeltals or Tzotzils from the Chiapas Highlands who used to come to earn money to pay for a '*mayordomía*', a stewardship in their village. Instead, the '*Chapines*', Indians from Guatemala, come to work in the plantations. This in the midst of the genocidal war that the Guatemalan army unleashed against its own fellow citizens, peoples whose languages were Kanjobal, Man Chuj, Jacalteco, Tojolabal, Quiché, Kakchiquel, and others. Trapped in a pincer between the army and the guerrillas, this movement led to the entry into Mexico of more than 150,000 Guatemalan refugees between 1978 and 1984.

They settled in the regions of Amparo Agua Tinta, Cuauhtémoc, the Montebello lakes, Ocosingo, and Las Margaritas. To prevent incursions against them by the '*kaibiles*' of the Guatemalan army, they were relocated some time later to the Mexican states of Campeche and Quintana Roo. The Indian refugees also came through the Lacandón rainforest, which led to a hasty policy of colonization of the border zone that has caused accelerated deforestation of the rainforest without creating long-term sustainable livelihoods for the settlers.

Many Guatemalans also came in through the southern border during the eighties and nineties, followed by other Central Americans, on their way towards '*el Norte*', the United States and its glittering hope of jobs. The prolonged civil war in El Salvador and the unfavourable economic conditions in Honduras, Nicaragua and Panama also pushed migrants north. It is as if the point of return of migrations from the north to the south fifteen thousand years ago has arrived, so that today the flow is reversed and intensified from south to north.

In the nineties, Central American migrants are sometimes joined by migrants from Colombia, Ecuador, Peru, Bolivia and other South American countries. As a consequence of the debt crisis and unfair conditions in international trade between the North and the South, it can be expected that this flow will continue to increase, since, if financial capital does not flow towards where the people live, then the people will no doubt continue to flow towards the places where financial capital can at least offer them economic survival.

Summarizing the above, in the last three decades of the twentieth century, Mexico has received a whole range of migrants from Latin American countries, from political exiles to economic refugees and survivors of military genocides, each group bringing with them their music and skills. The South American exiles who came to Mexico in the seventies brought with them Andean music, *sambas* and *milongas*, songs of protest, nonconformist political theories, and a feverish demand for a new distribution of power. The Guatemalan refugees brought, embroidered on their rich costumes, the memory of a great Mayan civilization and their complex history with Chiapas and Mexicans.

9.11 *Mestizo* Cultures in Movement

Mixed cultures are by definition liminal in the sense that they take from two or more cultures and create new intertwined cultural patterns. In the case of Mexico, the '*mestizo*' (mixed) cultures unite two historically antagonistic cultures, the indigenous and Spanish cultures, and, in turn, produce a thousand and one faces.

Debates about the history of Mexico have recognized the exceptional richness of *mestizo* cultures. They not only became prevalent all over the country but also configured distinctive regional cultures. For example, the culture of the *Jarochos* on the Gulf coast, through a mixture of indigenous, European, and African cultures, created exuberant '*sones*' (music played with guitars), graceful dances and ballads, and a festive form, the '*huapango*', also nourished by the rhythms of Caribbean music and Spanish poetry. A *Yucatecan* culture also flourished, a mixture of the arrogance of the 'Casta Divina', the 'Divine Caste' of the descendants of the Spanish, and the very rich Mayan cultural substrata, offering literature, music and dance characterized by the dazzling whiteness of the tropics. We are all very familiar also with the *Tapatia* culture, one in which, as the song claims, women have eyes that flutter like paper butterflies, Pedro Paramo's introspection erases the border of life and death, and the secretive spaces of its churches and convents hushes passions into silence. Other distinctive regional cultures may be identified in the North, the Isthmus of Tehuantepec, the Bajío and the Huasteca regions.

Moreover, other cultures from Europe and Africa added to the *mestizo* cultural richness. Recent studies have shown the canvas of cultural contributions brought, among others from the Bantu, Wolof, Hausa, and Ashanti cultures. Just as in the case of indigenous peoples, Africans who arrived in New Spain lost their distinct identities and were subsumed under the terms '*negro*', '*moreno*', and African. Beginning in the sixteenth century, they left traces in the Mexican phenotype, in the coasts of Veracruz and Tabasco, and the Costa Chica of the Pacific. They left an indelible mark in the rhythms and cadences of the coastal and Isthmus music, even in their instruments, such as the marimba, in the wooden carvings of masks and figures, and in the mysteries of their rituals and dances.

But there were other guests who brought their gifts to the Mexican cultural fiesta: the Americans and other European nationalities, the Chinese and Lebanese, and so many others. Some of them stayed within cultural communities that kept up their traditions, some of which passed into the Mexican culture. Most of them were voluntary migrants who decided to come to live in Mexico because it offered them something that they had not found in their places of origin.

Mention should also be made of the exiles. In the 1940s, very important contingents of Spanish Republicans arrived, whose contributions to the artistic, intellectual, and academic production of Mexico are well known. This includes anthropology, a field which has had very distinguished exiled teachers, such as Pedro Bosch-Gimpera, Juan Comas, José Luis Lorenzo, Santiago Genovés and others. After that, in the sixties and seventies the Brazilians, Uruguayans, Chileans, and Argentinians landed in Mexico, most of them professionals and academics who have also generously offered their work and cultural contributions.

Many of these migrants have stayed on in Mexico, integrated into new cultural whirlpools. So Mexico's position as a country of cultural crossways has been reaffirmed, from the West, the East, the North and the South. Thus, in the long count of the millennia, we find ourselves again in an era of *global crossings*.

9.12 Global Change and Interdisciplinary Anthropology

A fourth revolution is beginning at the end of this millennium, after the previous three revolutions. The Neolithic revolution, already mentioned, brought about agriculture and settled villages. In Europe, in the nineteenth century, the industrial-urban revolution which transformed food production and manufacturing also brought about a demographic revolution which accelerated population growth. These two revolutions spread to the rest of the world unevenly throughout the twentieth century.

At present, the fourth revolution, ushered in by economic globalization, telecommunications, biotechnology and microelectronics, needs to find its place in the new demand for *sustainable development*. A growing body of scientific research is showing that atmospheric and ecosystem balances must be kept and that the planet cannot continue to be loaded with non-biodegradable and toxic waste. A world in which there is already an imbalance between countries in terms of the distribution of world income, world trade, energy use and the importation of forest products, among many other traits, and within countries because of great disparities in income distribution.

Today, technological and social processes continue to alter the geophysical, chemical and biological systems of the earth, and the interactions between them. Some of these changes could reach the point of making the planet uninhabitable for human beings, particularly because many of these processes are occurring at a rate that is unprecedented in history. For example, since 1950 the demand for energy in the world has grown fourfold; the world population has doubled. Added together, these new, cumulative phenomena are resulting in processes that will affect *all the inhabitants of the planet*. That is what is meant by *global change*, which requires that action be taken in the following fields:

1. Preventing global warming, the effects of which are already causing natural disasters, causing the flooding of sea coasts and deltas, and altering the agricultural cultivation pattern of different regions. There has always been a greenhouse effect on the earth, but this time the increased use of fossil fuels (principally oil and carbon) produces large amounts of gases such as carbon dioxide, methane and nitrous oxide that are no longer absorbed sufficiently by oceans, forests and rainforests because of deforestation. These gases cause a greenhouse effect that is heating up the atmosphere. It is known that about 80 % of these gases are produced by industries in the countries of the North and 20 % by industrial and livestock activities in the countries of the South. This means that the countries of the North have to change their consumption patterns in order to lower their levels of use of energy and other natural resources, and that the countries of the South have to reformulate their development plans so that the type of industrialization and urbanization that they promote will be sustainable.

2. Halting the thinning of the ozone layer that has occurred over Antarctica and is also beginning to occur over the Arctic. This is due to the use of chlorofluorocarbons used in aerosols, refrigeration gases, and others. When only a few aerosols and refrigerators were used, these gases did not cause any harm; however, now they are produced all over the world, they are damaging the ozone layer of the atmosphere. So it is necessary to realize that it is the *scale* of usage that is causing these global biogeochemical effects.

   These two phenomena occur *in relation to the global biogeochemical system*, but there are others that occur *all over the earth*, such as the loss of biodiversity and the loss of cultivatable lands due to erosion, desertification, and acidification.

3. Reversing the loss of biodiversity. This has to do with the extinction of hundreds of species of plants and animals, some of which have not even been classified by science. This is due, on the one hand, to direct consumption—eating turtle eggs when this amphibian is almost extinct, or selling parrots from the Lacandón jungle for purely decorative effects in bars, restaurants or private homes–, and the increase in population with its increased consumer demands brings about the same problem of an increase in *scale* as above.

    Moreover, an additional problem is caused by the altering of interactions between the components of the ecosystems. That is, species of flora and fauna become extinct when the circuits of interdependence and the habitat in which they live are destroyed.

4. Preventing the loss of cultivatable lands. This is due principally to: (a) the erosion of agricultural lands due to deforestation, monocultures, and the expansion of human settlements, principally in Latin America, Asia, and southern Africa; (b) desertification in the southern region of Africa, and (c) acidification of the lands in North America, including north-western Mexico, and other regions of the world in which fertilizers and pesticides are improperly used.

5. Particularly in developed countries, population is also mentioned as a decisive factor for achieving equilibrium throughout the planet. As has already been said, the world's population has doubled in the last forty years (Turner/Clark/Kates/Richards/Mathews/Meyer 1990: 43). This growth has greatly accelerated in the past two centuries, as mortality fell in most regions of the world, and yet it will still be concentrated in the poorest countries of the world in the next century.

In Mexico, the historical variations in the population process in the Valley of Mexico—the mythical Anáhuac, again—from 1150 BCE onwards have been described by Manzanilla (1993). In spite of the decrease in the population with the fall of the great urban states of the Classical Era, and the decimation of the indigenous people in their last Meso-American cities with the Spanish invasion, as described by (Matos 1993), starting in 1643–1644 the population began to recover, slowly at first and then at an accelerated rate, until it reached its maximum historical growth of 5.10 % in 1970. In the last twenty years, this growth has decreased gradually. However, given the age structure of population, that is, the large number of young people of reproductive age already in the population pyramid, it is estimated that by the year 2000 the population of Mexico will be 104.5 million people, and that of Mexico City 35 million.

On a world scale, it is estimated that by the year 2010, there will be eight billion people in the world, the majority of them in countries in the urban South, many of them trapped in a vicious circle of poverty and population increase.

As the population continues to grow at a rapid rate, there is always a debate about whether sufficient food can be produced and enough jobs created; whether there will be enough natural resources, land, forests, rainforests, water, and oil. Although some authors believe that new technologies will be able to substitute or recycle these resources, the fact is that in most of the countries of the South the population continues to grow at a faster rate than the economy, and even if a more equitable distribution of capital and technologies were to be achieved so as to raise the standard of living, in the medium term poverty will not disappear.

Therefore, the world's societies have to learn again how to manage the use of natural resources so humanity can continue to live on as a species. It is equally urgent to renegotiate the coexistence of nations, societies and ethnic groups, and to harmonize population growth with a sustainable economic growth of the economies in order to be able to continue to live on as civilized beings.

9.13 The Future Depends on Us

Anthropology has always turned its gaze towards the past. But at this juncture, as we catch glimpses of new threats, we turn to look towards the future.

Today, for example, while sitting in Mexico City it is possible to know instantly what is happening in Beijing, or New Delhi; and it will not be long before the whole Library of Congress may be transmitted through computers from Washington DC to Mexico City in three minutes! This new instantaneity of news and the proliferation of communications and travel networks have already begun to change the forms of culture in Mexico as in the rest of the world.

The mirror of anthropology, which gives back to us the knowledge of millennia of human cultures, confronts us today, as almost never before in history, with the challenge of reconstructing our way of life in order to be able to carry on as cultural beings on this planet.

9.14 Movement and Uncertainty

In sum, there is no reason for us to be surprised by movement: it has always been an intrinsic part of human life and, particularly, Mexican life. Now, perhaps an exceptional era of stability and certainty in history is just ending and for some time we will have to get used to living with uncertainty. We must remember, however, that movement—the *ollin* of the Nahuas—opens up unimagined opportunities when it is combined with the cultural richness and the will to survive that have characterized Meso-American cultures, *mestizo* cultures, and in fact all cultures of the world for so many millennia.

References

Chávez Galindo, Ana María, 1987: *Migración, fecundidad y anticoncepción en Baja California* (Mexico City: UNAM).
CONAPO (Consejo Nacional de Población), 1988: *México Demográfico: Breviario 1988* (Mexico City: CONAPO).

De la Peña, Guillermo, 1993: "La antropología urbana y los estudios urbanos", in: Arizpe, Lourdes (Coord.): *Antropología Breve de México* (Mexico City: Academia de la Investigación Científica–CRIM-UNAM): 265–288.

Del Val, José Manuel, 1993: "El indigenismo", in: Arizpe, Lourdes (Coord.): *Antropología Breve de México* (Mexico City: Academia de la Investigación Científica–CRIM-UNAM): 245–264.

García-Bárcena, Joaquín, 1993: "Prehistoria, sedentarización y las primeras civilizaciones de Mesoamérica", in: Arizpe, Lourdes (Coord.): *Antropología Breve de México* (Mexico City: Academia de la Investigación Científica–CRIM-UNAM): 13–56.

Geertz, Clifford, 1973: *The Interpretation of Cultures* (New York: Basic Books).

González, de la Rocha Mercedes, 1993: "Respuestas domésticas, respuestas femeninas: la organización social de la pobreza y la reproducción", in: Arizpe, Lourdes (Coord.): *Antropología Breve de México* (Mexico City: Academia de la Investigación Científica–CRIM-UNAM): 319–342.

Huntington, Samuel, 2004: "The Hispanic Challenge", in: Foreign Policy (March 1), at: http://www.foreignpolicy.com/articles/2004/03/01/the_hispanic_challenge (13 June 2013).

Huntington, Samuel, 2005: *Who are we? The Challenges to America's National Identity* (New York: Simon & Schuster).

Kirchoff, Paul, 1943: "Mesoamérica: sus límites geográficos, composición étnica y caracteres culturales", in: *Acta Americana*, 1: 92–107.

Manzanilla, Linda, 1993: "Surgimiento de los centros urbanos en Mesoamérica", in: Arizpe, Lourdes (Coord.): *Antropología Breve de México* (Mexico City: Academia de la Investigación Científica–CRIM-UNAM): 57–82.

Matos, Moctezuma Eduardo, 1993: "Las sociedades tardías de Mesoamérica", in: Arizpe, Lourdes (Coord.): *Antropología Breve de México* (Mexico City: Academia de la Investigación Científica–CRIM-UNAM): 83–120.

O'Gorman, Edmundo, 1958: *La invención de América* (Mexico City: FCE).

Pick, James B.; Tellis, Glenda L.; Butler, Edgar W.; Pavgi, Suhas, 1990: "Determinantes socioeconómicos de migración en Méico", in: *Estudios Demográficos y Urbanos*, 5,1,13 (January–April): 61–101.

Quezada, Noemí, 1993: "Sexualidad, religión y magia", in: Arizpe, Lourdes (Coord.): *Antropología Breve de México* (Mexico City: Academia de la Investigación Científica–CRIM-UNAM): 121–146.

Serrano, Sánchez Carlos, 1993: "Bioantropología de la población mexicana", in: Arizpe, Lourdes (Coord.): *Antropología Breve de México* (Mexico City: Academia de la Investigación Científica–CRIM-UNAM): 147–164.

Tejera, Gaona Héctor, 1993: "La comunidad indígena y campesina de México", in: Arizpe, Lourdes (Coord.): *Antropología Breve de México* (Mexico City: Academia de la Investigación Científica–CRIM-UNAM): 189–214.

Torres, Mejía Patricia, 1993: "El México obrero industrial", in: Arizpe, Lourdes (Coord.): *Antropología Breve de México* (Mexico City: Academia de la Investigación Científica–CRIM-UNAM): 289–318.

Turner, Billie Lee II; Clark, William C.; Kates, Robert W.; Richards, John F.; Mathews, Jessica T.; Meyer, William B., 1990: *The Earth as Transformed by Human Action: Global and Regional Changes in the Biosphere over the Past 300 Years* (New York: Cambridge University Press—Clark University).

Valiñas, Coalla Leopoldo, 1993: "Las lenguas indígenas mexicanas: entre la comunidad y la nación", in: Arizpe, Lourdes (Coord.): *Antropología Breve de México* (Mexico City: Academia de la Investigación Científica–CRIM-UNAM): 165–188.

The disappearance of small landholder agriculture in Mexico. A maize field in Chalcatzingo, 2012. This photo is from the author's photo collection

Chapter 10
How to Restore Social Sustainability in Mexico

10.1 Introduction

Thirty-five thousand deaths in the last decade in Mexico cannot be explained just by reference to the 'War on Drugs' launched by the government.[1] Increased inequality, massive outmigration to the United States, joblessness (especially among young women and men) and soaring crime are some of its more overt expressions. To restore social sustainability in Mexico we need to understand the direct and indirect causes, as well as the triggering factors, in order to prevent their continuation in the future.

The surge of violence in murders, kidnappings, 'feminicides' (a hate crime against women at the hands of husbands, boyfriends or close kinsmen, as defined in Mexican law), people trafficking and other such crimes is frequently explained by reference to the immediate triggering factors: police corruption, poverty leading to rage as inequality soars, violence which spreads from border zones, low quality of education, political corruption, family disintegration and many other specific and interrelated factors. Yet this does not explain why Mexicans have become so violent. Many of these triggering factors already existed. Why have such crimes exploded since the year 2000? Why has Mexico become such a fertile ground for organized crime and why has drug-related violence led to unprecedented levels of violence against women and young people, with the spread of vicious and heartless methods of killing formerly unknown in Mexico?

This paper focuses on these violent trends from the perspective of social sustainability, defined as the capacity of a society to create and train citizens who are able to work, pursue knowledge, participate politically, and build convivial social and personal relationships (Arizpe 1989). A society is no longer sustainable where elites break political rules and where, as one shocked commentator stated: "…we are now detecting that the reason for kidnapping is not simply economic

---

[1] This text was originally presented at the Coloquio UNAM, 2010, in Mexico City, and is unpublished.

L. Arizpe, *Migration, Women and Social Development*,
SpringerBriefs on Pioneers in Science and Practice 11,
DOI: 10.1007/978-3-319-06572-4_10, © The Author(s) 2014

but to cause pain. There is a mode of operation in which kidnapped victims are tortured and assassinated. This is independent of whether the ransom is paid".[2]

In this context it is necessary to examine the way in which slow and very unequal economic growth and the fact that the 'democratic transition' promised by the neo-conservative governments is now at a standstill are bringing about a breakdown of public ethics and of social norms for living together. This is fuelled by the complicities of politicians and drug traders, and the acceptance of the breakdown of moral standards as in the case of Father Marcial Maciel's very powerful political intervention in Mexico and in the Vatican, despite practices of child abuse and corruption at the highest institutional and media levels. Social trends including increases in domestic violence, drug addiction especially among young people, and suicides in younger age groups and especially among young women all point to a political and social breakdown which urgently needs to be halted. The 'War on Drugs' initiated by President Calderon, who brought the army on to the streets to fight the drug cartels, only illustrated the traditional Spanish saying "*el camino al infierno esta sembrado de buenas intenciones*" ("the road to hell is filled with good intentions").

In this paper, I try to analyse the causal patterns of social breakdown at different levels and then I propose a few immediate measures which should be implemented to restore sustainable forms of living together in Mexico.

10.2 The Violence of Economic Inequality and Exclusion

In general, research on free market policies in Latin America has shown that few economic and social gains have been made (UNAM 2010; Birdsall/de la Torre/Menezes 2007). At the Committee for Development Policy of the UN Economic and Social Council, we confirmed this finding.[3] In countries where political structures have created a high level of economic inequality, such as Mexico, the rich have become much richer, and monopolies, duopolies and oligopolies have increased, there are high levels of unemployment, and rates of pay have fallen. This may be the necessary explanation for the surge in violence in Mexico, but it is not a sufficient explanation. I posit that a more adequate explanation has to do with the violent destruction of traditional Mexican agrarian society—in only twenty years—, the backlash against women's advances brought about by old patriarchal religious ideologies, the demolition of public education and its ideals of civic service (the neo-conservative governments have tried to remove courses on civics and philosophy in high schools), the blurring of political and national identities, and the weakness of the State in confronting the generalized impunity to corruption and crime. It is only following an understanding of these trends that the spirals of rage and despair which are now having devastating effects in Mexico can be stopped.

---

[2] "Entrevista a Alejandro Desfassiaux, presidente del CNSP", in: *La Jornada*, 7 December 2010: 11.

[3] See also UN Committee for Development Policy, UN Economic and Social Council, 2007–2010; at: www.un.org/en/development/desa/ (1 November 2013).

The data analysed in the following pages comes from the National Statistical Institute of Mexico (INEGI),[4] from recent studies by social scientists, and from direct observation during my fieldwork in many rural communities in central Mexico.

In the report *México frente a la crisis: hacia un nuevo curso de desarrollo* (*Mexico Facing the Crisis: Towards a New Development Path*, UNAM 2010), published by the National Autonomous University of Mexico, data show that the lack of development in Mexico comes as a result of adopting the policies of the Washington Consensus, as has been reported in other countries. In fact, in the words of Richard Jolly, former director of the Institute for Development Studies at the University of Sussex in the UK, the goals of the Washington Consensus were "...consistent with a neoliberal perspective of the economy but were not necessarily derived from them. Rather, they represented the interests and politics of industrial countries, especially the United States, the United Kingdom, Germany and Japan" (2010: 11–36). The interests of industrialized countries and private companies went further than that, to such an extent that, in his speech to the US Senate, Alan Greenspan, then President of the Federal Reserve, admitted: "I made a mistake in presuming that the self-interests of organizations, specifically banks and others, were such that they were the best capable of protecting their own shareholders and their equity in the firms".[5]

After 2000, in Mexico, economic growth was very low while all other indicators of well-being fell rapidly. The 2010 Competitiveness Report published by the Global Economic Forum of Davos states that "...in particular, the fact that the higher education and labour training institutions (of Mexico) is in Rank 79, seems to indicate that it does not produce the labour force nor the scientists nor engineers who could give a considerable push to technology and innovation" (WEF 2010). As the Rector of the National Autonomous University of Mexico, José Narro, has stated, Mexico ranks lowest among OECD countries in critical measures for the construction of the knowledge society. Data from the Economic Commission for Latin America and the Caribbean (ECLAC) have confirmed that Mexico is one of the few countries in the region in which poverty has not decreased, while on the other hand economic inequality has greatly increased. While the Gini coefficient decreased in Brazil, it actually rose in Mexico.

10.3   The Destruction of the Agrarian Sector in Mexico as a Direct Cause of the Expansion of the Drug Trade

After 1994, once the North American Free Trade Agreement (NAFTA) came into force and until 2005, Mexican government spending on the development of agriculture and fisheries dropped by 54.7 %, although there was a slight rise a few years later. By 2005 government loans for agriculture represented only 12.6 % of

---

[4] At: http://www.inegi.org.mx/ (1 November 2013).

[5] "Greenspan 'shocked' that free markets are flawed", in: *The New York Times*, 23 October 2008, at: http://www.nytimes.com/2008/10/23/business/worldbusiness/23iht-gspan.4.17206624.html?_r=0 (1 November 2013).

credits issued before 1994.[6] As had been foreseen, NAFTA policies drove smallholder agricultural households to bankruptcy. Whatever their economic contribution to the economy, such households created complex institutions of political and social services, including safety nets in uncertain times, which also provided strong identity and cooperation networks. Moreover, they represented the foundation of an ancestral history that gave Mexico great political flexibility and social resilience, and they were the source of a strong emblematic representation in world culture.

Since the nineties, faced with the bankruptcy of their rural households, young men and women had four options: wage labour in agriculture, migration to the United States, educational attainment for economic mobility, and the drug trade (for further analysis see Arizpe 2014). In the last ten years the first option has recalled the worst situations of indentured agricultural labourers around the world: health-breaking work, poverty, and brutal treatment. The workers' situation has deteriorated to such an extent that agribusiness companies, supported by the government, no longer allow anthropologists or any other outsiders to visit their agricultural fields (Carton/Francis/Lara 2008).

As to the second option, many rural families have made the effort to send one or two of their children to technical schools, pre-university schools or universities, including girls. Many of them have studied in areas of specialization that could be applied to rural areas such as agronomy, veterinary science, and nursing. Unfortunately, many others have found that no jobs were available in rural areas and in the cities they could not compete in the labour market with urban graduates. Large-scale private industries have been established in rural areas with the support of the government. For example, mainly in agribusiness for export, we find subsidiaries of global companies which have preferred to bring in middle-level employees from their home offices, creating no jobs locally. Small businesses, most of them set up by migrants in the service sector, are too small to create much employment. At the same time, the service sector and the informal sector in the cities has become saturated, so finding an income-generating activity has become ever more difficult. As a result unemployed young graduates with higher skills have been taking part in a constant flow of undocumented migration towards the United States and Canada.[7]

An aspect of this flow that has rarely been examined is that large numbers of young women of marriageable age (17–25) are left in a void. Since their families do not send them into higher education, they are left with nothing to do in their town or village, as their potential boyfriends have migrated. What happens then can be gleaned only superficially from our fieldwork data. A few of them become unruly and might try drugs or alcohol, especially if they live with their grandparents as their parents are in the US, and so they are 'ruined' as potential brides; others become depressed; yet others leave to work in unskilled jobs in the city; and

---

[6] Bank of Mexico (Banxico). Statistics available at: http://www.banxico.org.mx/publicaciones-y-discursos/index.html (1 November 2013).

[7] For example, in the town of Tlacotepec in the state of Morelos, 80 % of young boys leaving high school opted for undocumented migration to the US. Similar rates were found in other towns in the region (see Arizpe 2006).

others try to migrate to the US through the long chains of family and kin who are already established there. Is this kind of behaviour the explanation for the increase in teenage pregnancies, rates of depression and suicides among young women? To this one must add the pressure of the neo-conservative governments which maintain that the only role for a woman is being a mother, staying at home, and serving the family. Such pressures in situations where conditions do not allow women to comply with such values will only add anguish to an already stressful situation.

The third option, migrating to the United States, has become much more difficult with the stricter police control by the United States along its border. Whatever the effect of such restrictions on incoming migration flows to the US—increased also by the South and Central American migrants crossing Mexico towards 'el Norte'—, the migratory process has become a criminalized process, and its trajectory is fraught with dangers, from rape and physical violence to assassination or death by thirst in the desert. Such violence, in turn, has turned both police and migrants into much more violent people, entangled in a spiral of corruption, delinquency, and brutality that has permeated cities and migrant communities along the border and beyond.[8] There have been more than 700 'feminicides' along the border, especially centred in Ciudad Juarez, allegedly linked to 'snuff' movies, that neither the police, the government, nor the Mexican army have been able to stop.

So what options are left for young people expelled from their ancient agrarian society, discriminated against in education, excluded from jobs, and pushed towards illegal migration? In this heartbreaking collapse of economic and social sustainability in many regions of Mexico, only one business has thrived: the drug trade.

In brief, when the government stopped economic assistance and credit for agriculture for small farmers, it was drug traffickers who invested in their fields. When their crops failed, the drug traffickers gave them new credits, bought their agricultural inputs (either for traditional crops or for drug crops), and helped organize transport and marketing for their crops. Of course, they also provided very attractive jobs for their sons[9]—lots of money and women, power and respect, travel and guns. With the withdrawal of the government from the social services, the drug trade provided additional funding for families in case of accidents or death, health needs, communication with family members in the US, and so on. As the drug traders became patrons, they corrupted the local, then the regional, then the state governments in many regions. Concomitantly, the more big business set up barriers and refused to pay millions of pesos in taxes, the more the drug cartels stepped into the market, the social services, and the federal political system.

A vicious twist was that, as criminality exploded, and the arms industries on the US side of the border sold more and more arms to Mexican black operators, the

---

[8] A case in point is that of the Mara Salvatrucha. In the book *Los Retos Culturales de México* (Arizpe 2004), Ma. Elena Ramírez described how the Mara were a string of youth gangs along the migrant corridor between El Salvador and the US who played around with tattoos, bulging clothes and guns. Ten years later the Mara Salvatrucha is one of the most violent and brutal criminal gangs, feared by everyone and uncontrollable (Ramírez 2004).

[9] A phrase originally published in the journal *La Jornada* became famous: "Más vale cinco años como rey que cincuenta de buey", "Five years as a king is worth more than fifty as a mule".

drug cartels themselves offered to provide security and protection while effectively taking whole regions out of the control of the official federal police and army.

10.4 The Violence of Joblessness

According to a study by Ciro Murayama, in the first decade of this century Mexico went through a period of economic stagnation (Murayama 2006). During the government of Vicente Fox, the unemployment rate doubled from 1.60 % in 2000 to 3.20 % in 2006, and during the first half of Felipe Calderon's government, from 2006 to 2009, it increased to 5.17 %. Additionally, from 2003 to 2010, wages lost 51 % of their purchasing power (Murayama 2010: 71–85).

In a recent study, Samaniego (2010) has estimated that the number of unemployed in Mexico had increased to 2.5 million in 2010 (see also UNAM 2010). The author also refers to the worsening of labour conditions following the flexibilization of work contracts that has reduced access to social security, among other changes. Wages have also fallen as more workers are employed for less than a full day's work (Samaniego 2010; UNAM 2010).

Gendered markets had already led to differential impacts on the employment or women and men in Mexico even before present trends (Katz/Correia 2001). While there are more women than men in the jobs with lower wages and less productivity, the loss of jobs in the economic crisis has hit men more than women, as explained by research carried out by Saul Escobar. With higher unemployment among men, women are being pulled into wage employment: "The feminization of employment due to the crisis is not good news, because alongside this trend there is a trend towards employment becoming more precarious, that is, a loss of quality in paid work" (Escobar 2009: 79). Besides, more women than men accept a job without a written contract, whether temporary or permanent. This is a new vulnerability that affects women and the security of a stable income for their families.

Table 10.1 on women's economic participation in Mexico provides the basic statistical data of the economic trends currently affecting women in Mexico.

Another important indicator is that the percentage of female-headed households has increased from 17.8 % in 1995 to 23.1 % in 2010.[10] No studies exist which explain the link between the rise of male unemployment and increased domestic violence and abandonment of the family by men unable to fulfil their role as breadwinners, the outmigration of husbands and sons, or the death of males in the family. On this latter point, I would ask what nobody is asking: how many of the mostly male deaths attributed to the 'War on Drugs' left widows, mothers or sisters in destitution? Has the government bothered to look into this new situation and its consequences? How are all these women surviving: by migrating to the US, by going into prostitution or the grey 'entertainment' industry, by participating in kidnappings, or by becoming part of drug trafficking gangs?

---

[10] INMUJERES (National Institute for Women) "Sistema de Información Estadística para Mujeres y Hombres", at: http://estadistica.inmujeres.gob.mx/formas/index.php (1 November 2013).

Table 10.1   Women in the Mexican economy 1995–2010

Indicator	1995	2010
Economic participation rate (per cent)	W 34.5[a]	W 42.5[b]
	M 78.2	M 77.6
Percentage of working population that receives no income	W 17.5[a]	W 9.6[b]
	M 12.4	M 7.7
Percentage of working population that receives five times or more the minimum wage	W 4.7[a]	W 6.5[b]
	M 8.3	M 10.1

Sources [a]INEGI "Ocupación y Empleo", at: http://www.inegi.org.mx (1 November 2013); [b]INMUJERES "Sistema de Información Estadística para Mujeres y Hombres", at: http://estadistica. inmujeres.gob.mx/formas/index.php (1 November 2013)

10.5   The New Vulnerabilities of Women in Mexico

An important point is highlighted by Norma Samaniego: the fact that the economic crisis is having a devastating demographic effect. The 'demographic bonus', that is, the high proportion of young people present in the demographic pyramid, seen as a great opportunity for countries to receive a boost to their economic growth, in the case of Mexico has vanished. In the previous pages, our fieldwork data has showed how young men and women from rural areas were trapped in a jobless environment. Samaniego (2010: 50) stresses that the informal sector of the cities has been saturated and is no longer an escape valve. Many young people have gone back to rural areas but in the new 'rurality without agriculture', she says, they are unable to find employment.

The future of a country, it is well known, depends on education and training opportunities for the young. Once they grow up in unstable social environments, excluded from economic opportunities and paths of social mobility, it is extremely difficult to bring them back to stable jobs and on to stable social and personal tracks. These young people in Mexico are now called '*ni-nis*', an abbreviation for "*ni estudio ni trabajo*", "I neither study nor work", and their number has been estimated at around seven million.

The Secretary for Education of the neo-conservative government, Alonso Lujambio, in trying to minimize the number of unemployed young men and women, recently stated that young girls between the ages of 13 and 19 were not 'unemployed' because they are carrying out domestic chores! It is this kind of dismissal by the neo-conservative Partido Accion Nacional (PAN) government, publicly expressed towards women studying or working outside the home, which is creating such a dissonance in young women's views of themselves. Women are being urged by the PAN government to have more children. Young teenage girls who get pregnant receive government scholarships. Yet the strongest thrust in their policies is towards limiting abortion. The traditional voluntary birth control policies of previous PRI governments have been minimized, while PAN government hospitals and clinics are making it very difficult for women to obtain contraceptive and other reproductive services, as we were was able to observe in fieldwork research. Also, pro-life activists supported by the Felipe Calderon government have been able to have legislation approved in many states to limit women's access to abortion, even in cases of rape. None of these policies were publicly acknowledged except in declarations by Catholic archbishops. While implementing policies to restrict women's reproductive rights and sexuality, however, neither the neo-conservative government nor the Church in Mexico acknowledged the child abuse and corruption among the clergy, particularly that of one of the major political Catholic leaders, Father Marcial Maciel, who was strongly supported by the Vatican.

10.6   Recent Trends in the *Maquiladoras* and the Feminization of Mexican Migration to the United States

In the eighties, the '*maquiladora*' programme along the border, which in its second phase had increased female employment in assembly plants since the seventies, began substituting women workers for men. By January 1990, women made up 60.7 % of workers, but this figure had dropped to 53.8 % in 2005.[11] In a recent survey of migrant women workers in border towns, when asked about their situation in 2009 as compared to 1993, 62 % said it was 'worse', 24 % said it was 'better', and 15 % said it was 'the same'.[12] Most of the women workers who had been dismissed could not go back to their communities of origin and so stayed on in the border towns. Left with no support from any social service institution, the spiral of poverty, insecurity and violence has led to the historically unprecedented level of 'feminicides' along the border, which has spread to other regions of Mexico. With the constant traffic in migrants, drugs and illicit activities, anyone who grew up in such environments was drawn into this spiral of corruption and brutality. The assembly plants did not feel responsible, police became highly corrupt and government institutions were either deeply embedded in such activities or were totally incapable of handling such situations. The army, the police, the drug and people traffickers and organized crime have added to thousands of assassinations, in a blatant example of how deregulated capitalism, government impotence and the complicity of the powerful can lead to such tragedies.

The violence against women is embedded in the context of the 'production of violence' studied by Gutiérrez Guerrero (2010). The horizon described is extreme: the number of homicides linked to organized crime increased from 1,000 to 2,300 between 2001 and 2007. Later, following military intervention, the figures increased to 5,207 'executions'—purportedly between rival drug cartels—in 2008, 6,587 in 2009, and approximately 11,800 in 2010. Gutiérrez focuses narrowly on the

---

[11] INEGI (National Institute for Geography and Statistics), at: http://www.inegi.org.mx/sistemas/bie/ (1 November 2013).

[12] "Libre comercio y el trabajo: Trabajadores mexicanos dicen que TLCAN fue un engaño", in: *CFO: Comité Fronterizo de Obrer@s*, March 2009, at: http://www.cfomaquiladoras.org/ libre_comercio_ytrabajador.html (1 November 2013); see also Fleck (2001).

detonators of violence linked to the drug trade: places, poverty, quality of life and human development, gangs, and police inefficiency, especially in Ciudad Juárez.

As to Mexican women migrants, the National Women's Institute estimates that 2.4 million Mexican-born women now live in the United States. The Institute emphasizes "the extreme vulnerability in which the migrant population finds itself, especially women, girls and boys due to discrimination, the risks they are facing and the lack of access to their rights".[13] In the last fifteen years, the 'feminization' of international migration—which in fact applies only to some regions in the world—, has been linked to the growth of the 'care economy' in industrialized countries (Benería/Berik/Floro 2003). More Mexican women have migrated to work in the domestic area, as well as in nursing and care of the elderly. In our fieldwork research we found that it was especially women over the age of forty who have no option of local employment or whose migrant husbands had stopped sending back money who left for the United States. In most cases they leave their teenage children with their grandparents. In some cases this creates tensions as the grandparents do not have the authority to curb the adolescents' behaviour. In migration studies in Latin America, such situations are now called the 'care deficit' created by the outmigration of mothers, especially heads of households. This care deficit is even more serious in regions of Mexico where, as mentioned earlier, male deaths or imprisonment due to the 'War on Drugs' has left families in greater conditions of insecurity and destitution. In other words, today's poverty in Mexico is not the same as traditional poverty: it is now deeply tinged with blood, rage and social disorganization.

10.7 Conclusions

The dire panorama described illustrates diverse phenomena through which a social crisis becomes manifest. These phenomena are part of more general processes, and unless their elements are understood, the alarming risks we are currently facing will not be overcome. These interconnected elements that have not yet been thoroughly analysed enable us to respond, at least partially, to the question of why so many Mexicans have become so violent over the past few years and why violence has threatened so massively and brutally the basic principles of human dignity, well-being and physical integrity.

Following this preliminary analysis, the following actions are suggested in order to rebuild social sustainability in Mexico:

1. Change the departure point of development thinking. Mexico is not an economy, it is a society, and as such it requires an integral development policy comprising equilibrated economic and social guidelines, mediated by an efficient public administration that involves all actors in negotiating a shared development.
2. Support and deepen studies, as happens in most emerging countries, which may lead to proposals that enable negotiating the models and conditions of the International

---

[13] INMUJERES (National Women's Institute) Press release no. 102, 17 December 2010.

Monetary Fund and the World Bank. In particular, Mexico must adhere to the new trends within these institutions that provide advocacy for such readjustments.

3. Promote an official initiative and a deep negotiation process that will put a stop to the causes, not just the symptoms, of violence and social decay, which affect all levels of Mexican society. For this we must recognize that economic inequality and social injustice are forms of violence per se, which translate into different constellations of violent activities that, if unresolved, will accumulate and grow, leading to the dehumanization of social relations.

4. Tackle the causes of social decay with political action at the highest level. This means: (a) reconstructing the vocation of the State to serve all Mexicans, considering all economic groups as core development agents, not just those that amass economic privilege, and preventing public servants minimizing the severe problems that citizens face; and (b) recognizing that development implies two co-evolutionary processes: growth in economic production and the care of social reproduction, without which, as has become manifest during the last decade, all Mexicans are deprived of the benefits of development.

5. Moving from a moralist doctrinal vision that sets artificial boundaries to situations and conceives a punitive outcome for what it considers illicit conduct and actions to a political and scientific analysis that accounts for a constellation of factors that cause such conduct and actions, which are frequently a defence in the face of desperate situations and are clearly distinct from those that are markedly antisocial.

6. Reversing the abandonment of peasant families and communities by the governments over the last decade, which has led to an exponential growth in the support bases of drug traffickers, at the same time as ending the police inefficiency that has enabled organized crime to replace social security, policing, and State activities in many regions of the country.

7. Generating and disseminating a sense of responsibility surrounding the new vulnerabilities of women due to: (1) the loss of the purchasing power of wages, high female unemployment and the precariousness of employment; (2) the increase in family work due to mounting insecurity and the decrease in the provision of social services; (3) the increase in violence against women by their spouses and the impunity of feminicides made possible by complicity between government and organized crime at the highest level. Following this last point, it is crucial to revise neo-conservative policies that encourage discrimination against women on all fronts and lead to their precariousness of employment, increase their defencelessness in the face of increasing violence, and bring about their moral anguish as they are unable to fulfil the demands of obsolete doctrinal values and demands in the context of the world today.

8. Reconstructing a vision of State and citizenship that endorses the concept of belonging to the same society for all citizens, that effectively guarantees freedom and individual liberties, and that provides a collective reference framework and future project for all, especially young people. In parallel, creating a new perception of politics, according to which different groups of citizens can disagree at the same time as they share equally a stake in the pathway for the nation and the world.

Finally, economic inequality and exclusion will lead to further violence. These veiled forms of violence, justified by policies that create economic and political monopolies and oligopolies, have led to an unprecedented loss of social sustainability in Mexico, resulting in physical violence and dehumanization. It is important to reconstruct a vision of justice and equality that is opposed to the particular interests of an ever-richer elite, to discrimination disguised as religious doctrine, and to the defencelessness of consumers and citizens. It is an urgent matter to bring about this vision and the politics of development in order to restore the pathway to a negotiated and shared future.

References

Arizpe, Lourdes, 1989: "On the Social and Cultural Sustainability of World Development", in: Emmerij, Louis (Ed.): *One World or Several?* (Paris: OECD Development Centre): 207–219.

Arizpe, Lourdes (Ed.), 2004: *Los retos culturales de México* (Mexico City: Cámara de Senadores–CRIM—UNAM–Miguel Ángel Porrúa).

Arizpe, Lourdes (Ed.), 2006: *Los retos culturales de México frente a la globalización* (Mexico City: Cámara de Diputados–CRIM-UNAM–Miguel Ángel Porrúa).

Arizpe, Lourdes, 2014: *Lourdes Arizpe: A Mexican Pioneer in Anthropology* (Berlin: Springer).

UNAM (Universidad Nacional Autónoma de México), 2010: *México frente a la crisis: Hacia un Nuevo curso de desarrollo* (Mexico City: UNAM).

Benería, Lourdes; Berik, Gunseli; Floro, Maria, 2003: *Gender, Development and Globalization: Economics as if All People Mattered* (New York: Routledge).

Birdsall, Nancu; de la Torre, Augusto; Menezes, Rachel, 2007: *Fair Growth: Economic Policies for Latin America's Poor and Middle-Income Majority* (Washington: Center for Global Development).

Carton de Grammont Bernet, Hubert; Francis, Marie; Lara Flores, Sara María, 2008: *Encuesta a hogares de jornaleros igrantes en regiones hortícolas de México* (Mexico City: IIS-UNAM).

Escobar Toledo, Saúl, 2009: "El empleo en 2009: breve recuento de daños", in: *Economía UNAM*, 7,20: 79–85.

Fleck, Susan, 2001: "A gender perspective on Maquila employment and wages in Mexico", in: Katz, Elizabeth G.; Correia, Maria C. (Eds.): *The economics of gender in Mexico: work, family, state, and market* (Washington: The International Bank for Reconstruction and Development): 133–173.

Gutiérrez Guerrero, Eduardo, 2010: "¿Cómo reducir la violencia en México?", in: *Nexos*, 395 (November): 24–33.

Jolly, Richard, 2010: "Employment, Basic Needs and Human Development: Elements for a New International Paradigm in Response to Crisis", in: *Journal of Human Development and Capabilities*, 11,1: 11–36.

Katz, Elizabeth G.; Correia, Maria C., 2001: *The Economics of Gender in Mexico: Work, Family, State and Market* (Washington DC: World Bank).

Murayama, Ciro, 2006: "México 2000–2006: la economía estancada", in: Sánchez Rebolledo, Adolfo (Coord.): *¿Qué país nos deja Fox? Los claroscuros del gobierno del cambio* (Mexico City: Norma): 105–132.

Murayama, Ciro, 2010: "Juventud y crisis: ¿hacia una generación perdida?, in: *Economía UNAM*, 7,20: 71–78.

Ramírez Parra, Ma. Eugenia, 2004: "Maras salvatruchas, nuevas culturas en la frontera sur", in: Arizpe, Lourdes (Ed.), 2004: *Los retos culturales de México* (Mexico City: Cámara de Senadores–CRIM—UNAM–Miguel Ángel Porrúa): 67–72.

Samaniego, Norma, 2010: "El Empleo y la Crisis. Precarización y Nuevas Válvulas de Escape", in: *Economía UNAM*, 7,20: 47–70.

WEF (World Economic Forum), 2010: *The Global Competitiveness Report 2010–11* (Geneva: World Economic Forum).

Lourdes Arizpe's first godchild, daughter of a young Nahua couple in Zacatipan, 1970. This photo is from the author's photo collection

Chapter 11
The Social Dimensions of Population

Lourdes Arizpe and Margarita Velázquez

11.1 Introduction

While the scientific understanding of environmental and demographic change, as studied separately, is increasing dramatically, our ability to link the two in any synthetic and holistic manner lags behind. The central argument of this paper will be that the scientific community cannot use current models and methodologies for understanding the dynamic relationship between population and environment, but needs a new framework. This new framework will need to extend key definitions of issues and concepts and propose new methods for researching them. Population, for example, cannot be limited to population size, density, rate of increase, age distribution and sex ratios, but must also include access to resources, livelihoods, social dimensions of gender, and structures of power. New models have to be explored in which population control is not simply a question of family planning but of social and political planning (UN 1990: 202–216; Jacobson 1987: 152–54) in which the wasteful use of resources is not simply a question of finding new substitutes but of reshaping affluent life-styles (Meadows 1988: 332–349; Repetto 1987) and in which pollution control is not simply a matter of 'polluter pays' but also of emission controls, which in turn are associated with political and social processes. These will need to be models in which sustainability is seen not only as a global aggregate process but one that incorporates the policy goals of sustainable livelihoods for a majority of local peoples.

This text was originally published co-authored with Margarita Velázquez as: "The Social Dimensions of Population", in: Arizpe, Lourdes; Stone, Priscilla; Major, David C. (Eds.): *Population and environment: rethinking the debate* (Boulder: Westview, 1994): 15–40. The permission was granted on 12 November 2013 by Melissa Malone, Perseus Book Group, the legal successor of Westview Press.

L. Arizpe, *Migration, Women and Social Development*,
SpringerBriefs on Pioneers in Science and Practice 11,
DOI: 10.1007/978-3-319-06572-4_11, © The Author(s) 2014

This is no small undertaking, and yet the theoretical and empirical challenges posed by global environmental change are in themselves a whole new order of magnitude. Although at a global level many recognize the challenge of harmonizing population growth and human expectations with the rate at which the planet's natural resources are being used or polluted, we lack the models with which to understand and plan for these changes. Human control of the environment is being overridden by unexpected new phenomena—the greenhouse effect, leading to climate change, and ozone depletion—or by the cumulative effects of old phenomena—desertification, loss of biological and cultural diversity, and soil erosion, among others. Humans are vulnerable now to natural and human-made hazards of a different order than ever before.

Three factors distinguish what we face today from challenges of the past. First, the scale of such phenomena is much larger and the number of people who will be affected by these changes is historically unprecedented. Second, while ecological mismanagement did occur in the past, populations could opt for outmigration. Now, however, there is nowhere left to go. Third, the natural inequities in the geographical distribution of resources have been further aggravated by the concentration of human-made capital in industrialized nations and in elite circles of less developed countries.

A challenge such as this, of a higher magnitude and complexity than humanity has had to face in the past, requires conceptualization and planning at a more inclusive and complex level. But we lack the appropriate scientific and political frameworks; issues tend to be constructed, and dealt with, around single factor explanations and ensuing simplistic actions. Believing, for example, that population is the key cause of environmental degradation is a reductionist argument that leads to narrowly conceived policies. The complexity of the Issues involved actually requires a debate on the political and economic planning for a global world.

This chapter contends that the population-environment debate has become deadlocked because it has become a question of taking sides instead of delving deeply into the complexity of the issues. Also the tendency to use mechanistic, predictive models is inappropriate given the level of uncertainty. Population issues have been decontextualised from actual social environments as well as from the broader and more profound issues concerning the new, emerging economic and political structure of the world and its relationship to the resource base of the planet.

Some may argue that engaging in the analysis of such broad issues may distract from the urgent need to act on population problems. Experience shows, however, that policy solutions focused exclusively on deterring population growth in the short term are ineffective when compared to more encompassing economic and social reforms. The most urgent task, then, is to establish a hierarchy of goals—economic, ecological, social, and cultural—to better direct the already existing potential for action.

At present, the disarray in the debate on population and resource use has been attributed to the lack of reliable data and the uncertainty of predictions. But it is also associated with the failure to analyze population trends *in relationship to other processes*. This chapter contends, accordingly, that all demographic transitions have been embedded in broader socioeconomic transitions; that population growth is not a *driving* force but an *accelerating* force except under rare circumstances where all other conditions remain static; and that population growth can only be understood by analyzing it in relation to rates of growth in the consumption of natural and human-made resources. Finally, we argue, as many others have, that curbing population growth can only occur in the long term with plans for *sustainable* development at a national, regional and global scale (Ehrlich/Daily/Ehrlich/Matson/Vitousek 1989; Ehrlich/Erhlich 1991; Keyfitz 1991a, b; Costanza 1991; Leff 1990; Little/Horowitz/Nyerges 1987; Toledo 1990; Maihold/Urquidi 1990).

11.2 Population Trends at the Threshold of the New Millennium

The demographic transitions in North America and Western Europe at the end of the nineteenth century were linked to improved medical services and nutrition levels, which led to the decline in mortality due to infectious diseases (Demeny 1990; Lutz/Prinz 1991). But they were made possible by a number of interrelated changes, including the shift from an agricultural to an urban-industrial society and by associated changes in family composition, age at marriage, and education.

In contrast, mortality decline in less developed countries in the second half of the twentieth century has come about mainly as a result of improved medical and health care services in many cases without the accompanying social, economic, and political transformations. Since these socioeconomic transitions have occurred unevenly, frequently inequitably, and sometimes have even been reversed, the demographic transitions in such countries have not been completed, especially in Africa.

Some authors believe a general demographic transition is already underway. Mian Simon (1990), for example, argues that fertility rates have decreased in countries all over the world. Others reject this optimistic view or believe that such transitions are occurring too slowly (Ehrlich/Daily/Ehrlich/Matson/Vitousek 1989; Ehrlich/Ehrlich 1991; Grant/Tanton 1981). Recent figures, in fact, show that worldwide the crude birth rate decreased from 33.9 (1950–1970) to 27.1 (1985–1990), while the total fertility rate fell from 5.9 to 3.3 during that same period (World Resources Institute 1990: 256; see Table 11.1).

Lutz and Prinz state that projections for the next 30 years are actually rather reliable, since they are insensitive to minor changes in mortality, migration, and fertility (1991). In Fig. 11.1, they summarize projections according to different scenarios. It is estimated that world population will reach around 8 billion by the year 2010 (UNPF 1991: 3, 48; UNDESA 1989; Demeny 1990: 41; Sánchez/Castillejos/Rojas Bracho 1989: 16; UNDP 1990: 166).

Table 11.1 Fertility trends in the developing world

Year	Average no. births/woman
1950–1955	6.1
1955–1960	6.0
1960–1965	6.1
1965–1970	6.0
1970–1975	5.4
1975–1980	4.5
1980–1985	4.2

Source Bongaarts/Mauldin/Phillips (1990)

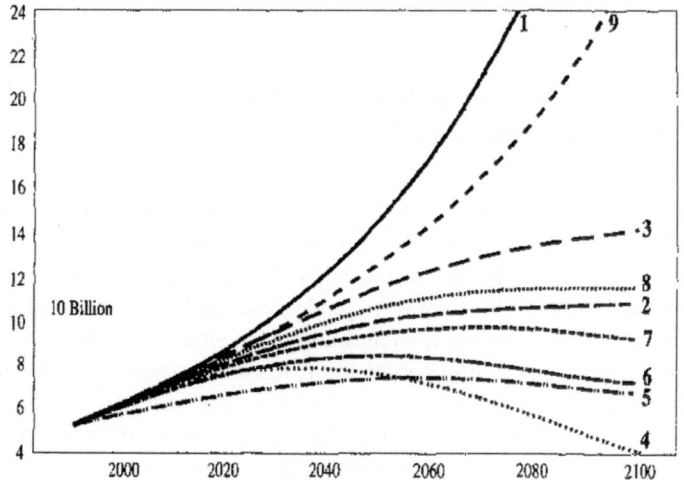

Scenarios

1: Constant Rates; constant 1985-1990 fertility and mortality rates
2: UN Medium Variant; strong fertility and mortality decline until 2025, then constant
3: Slow Fertility Decline; UN fertility decline 25 years delayed, UN medium mortality
4: Rapid Fertility Decline; TFR=1.4 all over the world in 2025, UN medium mortality
5: Immediate Replacement Fertility; assumed TFR=2.1 in 1990; UN medium mortality
6: Constant Mortality; TFR=2.1 all over the world in 2025, constant mortality
7: Slow Mortality Decline; UN mortality decline 25 years delayed, TFR=2.1 in 2025 8: Rapid
Mortality Decline; life expectancy of 80/85 years and TFR=2.1 in 2025
9: Third World Crisis; constant fertility and 10% increase in mortality in Africa and
Southern Asia; TFR=2.1 in 2025 and UN mortality for the rest of the world
Note: "TFR" is the Total Fertility Rate (= average number of children per woman)
Source: Lutz and Prinz 1991

Fig. 11.1 Total projected world population 1990–2100 according to scenario

As Lutz and Prinz (1991) emphasize, the *momentum* of population growth
has to be taken into account—the age structure of a fast growing population is so
young that even if fertility per woman declined to a very low level, the increasing number of young women entering reproductive ages will cause the population

to grow further for quite some time. For the year 2050 and beyond, projections begin to vary from 8 to 14 billion and they diverge even more widely for the next century after that (Lutz/Prinz 1991).

11.3 What Do Population Numbers Mean?

The concept of population as numbers of human bodies is of very limited use in understanding the future of societies in a global context. It is what these bodies do, what they extract and give back to the environment, what use they make of land, trees, and water, and what impact their commerce and industry have on their social and ecological systems that are crucial (Demeny 1988: 217; Harrison 1990; Durning 1991).

An early attempt to establish such a link was by estimating the planet's carrying capacity. Approximations varied widely, ranging from 7.5 billion (Gilland 1983), 12 billion (Clark 1958), 40 billion (Revelle 1976), to 50 billion (Brown 1954). The underlying problem is, of course, how to establish the appropriate level of kilocalories for each human being (Blaxter 1986). "For humans, a physical definition of needs may be irrelevant. Human needs and aspirations are culturally determined: they can and do grow so as to encompass an increasing amount of 'goods,' well beyond what is necessary for mere survival" (Demeny 1988: 215–216).

Other authors point out that the concept of carrying capacity and self-regulating mechanisms applied to animal populations should not be extrapolated to human populations (Sánchez/Castillejos/Rojas Bracho 1989: 26), and argue that socioeconomic, technological, or environmental changes are so decisive in altering carrying capacity that the concept itself is of little use (Blaikie/Brookfield 1987). A more appropriate exercise, then, should be to try to develop global accounting systems that relate population, per capita resource use, and wealth distribution.

A different approach is taken by those trying to forecast the socioeconomic and environmental impacts of population projections. Gordon and Suzuki, in their 1991 assessment of environmental change, present a harsh scenario for the year 2040: overpopulation, unbreathable air, high temperatures, desertification, loss of land due to soil erosion and rising seas, food rioting in the Third World, and so on. They insist that the 'sacred truths' that the earth is infinite and progress is possible must be discarded.

Another issue is the differential demographic growth of regions and of diverse ethnic and religious groups within countries. Here, population overlaps with political processes. Figure 11.2 shows the difference in the proportion of population between the North and the South based on United Nations 'medium' growth assumptions (UNDP 1990). By the year 2010, 6 out of every 7 people will live or will have been born in less developed countries of the South (FAO 1990; Table 11.2). Such differential growth is being seen increasingly as a potential threat to countries in the North (Grant/Tanton 1981).

During the last few years, population growth has increasingly been considered the driving force in environmental degradation, increased pollution, and in fostering international conflicts—for example, over water in the Middle East (Myers 1987). It is interesting to note that in linking population growth to environmental depletion,

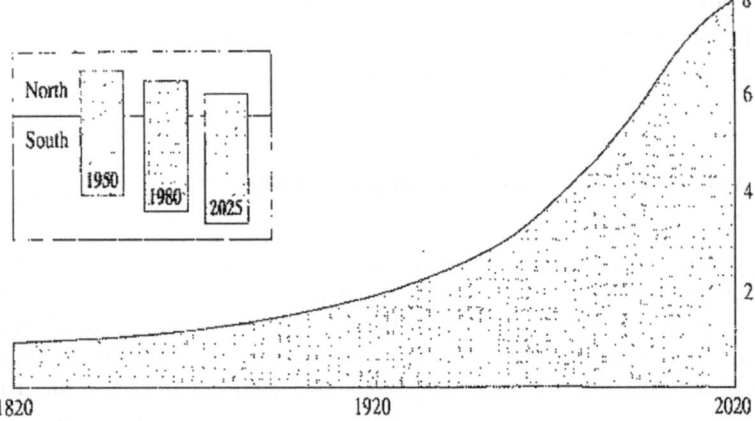

Fig. 11.2 World population trend and north south distribution (billions of people). *Source* United Nations Development Programme (1990)

Table 11.2 Population indicators for major regions of the world

Region	Population millions 1990	Av. rate growth % 1990–1995	IMR[a] per 1,000 1990	% Urban 1990	Urban growth % 1990–1995
World	5,292.2	1.7	63	45	3.0
'North'	1,206.6	0.5	12	73	0.8
'South'	4,084.6	2.1	70	37	4.2
Africa	642.1	3.0	94	34	4.9
North America	275.9	0.7	8	75	1.0
Latin America	448.1	1.9	48	72	2.6
Asia	3,112.7	1.8	64	34	4.2
Europe	498.4	0.2	11	73	0.7
Oceania	26.5	1.4	23	71	1.4
USSR	288.6	0.7	20	66	0.9

[a]Infant Mortality Rate
Source Sage/Redclift (1991)

the disparities in the natural resource base and in the distribution of goods and services in different societies are frequently left out of the picture. It would seem, in fact, that population growth, in some cases, is used to compensate for such existing disparities and inequities. To understand this, population numbers must be analyzed according to human development indicators. We will return to this point later.

11.4 The Population Debate

The debate on population has polarized into two major positions. One position holds that an increasing population is the principal driving force that threatens the planet's finite resources. Paul Ehrlich mentions that over consumption, increasing dependence on ecologically unsound technologies to supply that consumption, and unequal access to resources and poverty play a major role in environmental crises (Ehrlich/Daily/Ehrlich/Matson/Vitousek 1989). He concludes that "the key to understanding overpopulation is not population density but the number of people in an area relative to its resources and the capacity of the environment to sustain human activities" (Ehrlich/Ehrlich 1991: 38–39). Much more extreme views on this position have likened human population expansion to a cancerous growth bound to kill its hospitable planet (Hem 1990: 30) or to a jar of yeast, which makes alcohol that eventually kills it (Grant/Tanton 1981).

However, case studies have been reported indicating that "there is no linear relation between growing population and density, and such pressures [towards land degradation and desertification]" (Caldwell 1984). In fact, one study found that land degradation can occur under rising pressure of population on resources (PPR), under declining PPR, and without PPR (Blaikie/Brookfield 1987). Therefore, the scientific agenda must look toward more complex, systemic models where the effects of population pressures can be analyzed in relationship with other factors (García 1990). This would allow us to differentiate population as a 'proximate' cause of environmental degradation, from the concatenation of effects of population with other factors as the 'ultimate' cause of such degradation (Asian Development Bank 1991). Also concerned about population growth, other authors place greater emphasis on improving humanity's lot (Eckholm 1982).

The other position holds that population growth can be dealt with through technological solutions that human creativity will continue to find (Simon 1990; Kasun 1988). An even more optimistic view, originally put forth by Hirschman, saw high rates of population growth as stimulating economic development through inducing technological and organizational changes (1958).

This position, though, ignores the dangers of environmental depletion implicit in unchecked economic growth: consumption increases and rapidly growing populations that can put a very real burden upon the resources of the earth and bring about social and political strife for control of such resources. This position also makes the questionable assumption that technological creativity will have the same outcomes in the South as in the North. Finally, it heavily discounts the importance of the loss of biodiversity—a loss that is irreversible and whose human consequences are as yet unknown.

Neither position, according to some authors, represents the state of the art of scientific understanding (Johnson/Lee 1987; Repetto 1987). Other authors state that "population is not a relevant variable" in terms of resource depletion, and stress income inequality, that is, poverty, as a more important factor (Gallopin 1990; Leff 1990). More specifically, resource consumption, particularly over-consumption by the affluent, is considered by many authors as the key factor to environmental depletion (Hardoy/Satterthwatte 1991; Harrison 1990; Durning 1991). OECD countries represent only 16 % of the world's population and 24 % of land areas; but their economies account for about 72 % of world gross product, 78 % of road vehicles, and 50 % of global energy use. They generate about 76 % of world trade, 73 % of chemical products export, and 73 % of forest product imports (OECD 1991). The main short-term policy instrument in this case is reducing consumption.

Finally, a 'revisionist' position states that "neither alarmism nor total complacency about population growth can be supported by the current evidence" (Kelley 1986: 563–567). This author cites Kuznets' summary judgment: "we have not tested, or even approximated, empirical coefficients, with which to weigh the various positive and negative aspects of population growth. While we may be able to distinguish the advantages and disadvantages, we rarely know the character of the function that relates them to different magnitudes of population growth."

Importantly, historical demographic studies have demonstrated that no simple correlations can be established between population and environmental transformations. In the volume *The Earth Transformed by Human Action*, researchers found that the time scale of population variability is asynchronous with given environmental transformations and recovery (Whitmore/Turner/Johnson/Kates/Gottschang 1990: 37). Consistently, the authors stress "the need for caution in using population as a simple surrogate for environmental transformation" (Whitmore/Turner/Johnson/Kates/Gottschang 1990: 37).

They also found evidence of divergence between global and regional population trends and concluded that "if the experience of past regional population changes and their accompanying environmental transformations has relevance for the future, the projected global scale population 'levelling out' need not diminish the scale and profundity of global environmental change. This is particularly true on the regional or local scale, where global zero population growth (of population or transformation) need not be accompanied by local or regional equilibrium" (Whitmore/Turner/Johnson/Kates/Gottschang 1990: 37). Furthermore, they warn that "regional population declines are possible, potentially brutal, and even likely, accompaniments to zero population growth on a global scale" (Whitmore/Turner/Johnson/Kates/Gottschang 1990: 37).

Beyond a bipolar population debate, then, we must develop a clearer understanding of what population numbers mean in different social settings.

11.5 What the Population Numbers Mean: Migration and Urbanization

While intensive ethnographic research methods may point to unexpected and crucial relationships, the rethinking of questions must also extend into new areas of research. While most of the theorizing around the population-environment issue has been directed toward population growth, other demographic processes such as migration and urbanization must be clearly incorporated. Let us for illustrative purposes look at some of the macro issues surrounding these processes.

The demise of agrarian societies the world over, coupled with population increase, has contributed to massive migratory movements in the last few decades. In the early 1980s the number of economic migrants was estimated at around

20 million; adding a similar count for illegal migrants, perhaps 40–50 million people have moved in the hope of having a bigger share of the world's development benefits (UNDP 1990: 28). It is estimated that in the first decades of the next century such movements will increase and diversify, adding to their contingents ecological refugees, who will be moving principally from South to North, but also in regions within the South (UNDP 1990: 28).

The classic pattern of rural outmigration linked to price changes that led to the breakup of the European peasant economies—which sent out 52 million Central and Western European migrants overseas from 1848 to 1912, even though population growth was slow (Brinley 1961)—continues to be the main driving force in outmigration in many developing countries. At present, it is associated with the fall in the price of agricultural products in the world market: in the 1980s, subsidies in the European Economic Community and other developed economies to protect their own farmers have led to the breakup of peasant economies of many countries of the South, thus leading to massive rural exodus. Such pressures may also push farmers toward environmentally unsound activities: deforestation, monoculture, crop intensification, and overuse of fertilizers or pesticides, all of which mean depletion of natural resources.

Population growth may increase the rate at which such activities are carried out, but the structures that drive such activities are economic and financial, not demographic. If this were not the case, Europe would not have sent out as many migrants overseas during the last century.

The important question is what will happen as market forces continue to integrate agrarian societies into a globalized market, thus continuing pressures for rural outmigration. Measures to protect farmers from the market have failed in most countries. They have often transferred the problem to other countries, as has already been mentioned for the EEC, so that a local Western European solution to an economic and political problem has fuelled local problems of impoverishment, outmigration, and resource depletion in many countries of the South. In turn, as in the case of deforestation, that intensifies yet another *global* problem.

Traditional agrarian societies have responded in myriad ways, such as increased self-exploitation in work, decrease of caloric intake, especially among poor rural women, intensification and diversification of income-generating activities, and permanent or recurrent—'swallow' migration as it is called in Mexico—outmigration (Arizpe 1978, 1982).

An important question is whether having more children survive has been used by farmers in this unequal struggle to adapt to a global market. The answer is yes, in many ways; we focus on those related to migration strategies. Since farmers in traditional agrarian societies are not having more offspring but rather are finding that more of them survive, what is being asked of them is that they change an age-old pattern of reproduction. This requires not only a strong economic incentive, which in the case of the more than 1 billion rural poor in the South is not there, but a cultural and political change whereby rural people will feel they have a stake in the new global society that is being created. For small agricultural family producers, it is well known that children do have an economic value in farm

Table 11.3 Projected increases in urban population in major world regions, 1985–2000

Region	Urban population (millions)		Absolute increase (millions)	Percentage increase
	1985	2000		
Africa	174	361	187	108
Asia	700	1,187	487	70
Latin America	279	417	138	49
Oceania	1.3	2.3	1	77
Developing countries	1,154	1,967	813	70
Industrial countries	844	950	106	13
World	1,998	2,917	919	46

Source United Nations Development Programme (1990)

labour. Additionally, in recent decades, they have been helpful to their families in relay migration, by sending remittances back to offset the family economic deficit (Arizpe 1982). Such a strategy, though, is possible only when cities offer unlimited opportunities for economic gain, which increasingly is not the case in many mega-cities in developing countries.

Thus, adaptive strategies to disappearing rural livelihoods, given present market conditions, are limited, so it may be assumed that rural outmigration will continue to increase and to be directed toward mega-cities and to international destinations.

Rural-urban disparities continue to increase the attraction of cities: in most countries, urban incomes per person run 50–100 % higher than rural incomes (UNDP 1990: 30). In Nigeria the average urban family income in 1978–1979 was 4.6 times the rural; in Mexico urban per capita income was 2.6 times the rural (UNDP 1990: 30).

Projections show that urbanization will become the dominant social process in the next 50 years and most mega-cities wall be in low latitudes in tropical regions (Douglas 1991), as can be seen in Table 11.3. Some authors, however, believe that the rate of urbanization in the Third World will not be as high as United Nations figures suggest (Hardoy/Satterthwaite 1991).

The trend towards urbanization fostered by rural push factors rather than urban attraction factors is cause for concern, since "in the places where man's activities are most densely concentrated—his settlements—the environmental impact is greatest and the risks of environmental damage are most acute" (UN 1974). Indeed, since the United Nations pointed this out in 1974, urbanization has continued at a rapid pace. Globally, it has been forecasted that 24 million hectares of cropland will be transformed to urban-industrial uses by the year 2000; this is only 2 % of the world total, but it is equivalent to the present-day food supply of some 84 million people (Douglas 1991: 8). The loss of agricultural land to urbanization is most severe in the developing countries, where more than 476,000 ha of land a year will be built up in the remaining years of the twentieth century (World Resources Institute 1988, cited in Douglas 1991).

While in previous decades cities in developing countries were able to absorb, even under dire poverty conditions, migrants from their rural hinterland, this is no longer the case in the 1990s. Migrants from Eastern Europe and Africa are overflowing into Western Europe, and Peruvians and Bolivians have now joined Mexicans and Central Americans in migrating to the United States (Grant/Tanton 1981). No doubt the numbers of economic and ecological migrants knocking at the North's door would be lower if population growth in the South were decreased, but the trend would still be there. If capital investments do not flow to where the people are, then the people will flow to where capital investments are. One example will suffice: in California alone, it is estimated that 7 million jobs will be created in the 1990s, most of them low-paying, nonskilled jobs Americans will not be able to fill.

Consequently, with both push factors in the South and pull factors in the North at work, the flow of migrants from South to North will continue to grow in the next decades and may become one of the most contentious issues internationally, although an increase of South–South migration can also be foreseen. Thus, the population and environment debate must incorporate these concerns to better model these complex changes. This means, among other things, developing new research priorities and methods to monitor population processes but, in particular, analyzing them in relation to global change phenomena.

11.6 Exploring New Research Methods

Research should focus not on population as an isolated variable but on the relationship between population and the use of natural and human-made resources (Demeny 1988: 217; Harrison 1990; Durning 1991; Arizpe/Constanza/Lutz 1992). The methodological and theoretical challenges are considerable.

One research priority must be to explore methods for more precisely estimating the relationship between population and resource use. William Clark (1991) suggests that the 'Ehrlich identity' (Pollution/Area = People/Area × Economic Production/People × Pollution/Economic Production) can be operationalized as (CO_2 Emissions/km^2 = Population/km^2 × GNP/Population × CO_2 Emissions/GNP). Clark and his colleagues examined data for 12 countries from 1925 to 1985 and concluded that the same loading of pollution on the environment can come from radically different combinations of population size, consumption, and production. Thus no single factor dominates the changing patterns of environmental loadings across time.

During fieldwork in Zacualpan 2009 with Edith Pérez Flores. This photo was taken by Klaus Jacklein and is reproduced with the author's permission

Another research priority is to look at the effect that adding a new person has on resources according to consumption levels and the effect that efficiency has on rising levels of consumption. Kolsrud and Boyle (1993) examine population growth and energy efficiency in several countries and conclude that the very small population growth forecast for developed countries over the next 40 years will add a burden of CO_2 emissions that will be *equal* to that added by the much larger population growth forecast for the less developed countries. Improving energy efficiency in developed countries could dramatically decrease CO_2 emissions globally (if consumption per person remains constant). It is only under a scenario of severe constraints on emissions in the developed countries that population growth in less developed ones plays a major global role in emissions. If energy efficiency could be improved in the latter as well as the former, then population increase would play a much smaller role. José (1993), a participant at the meetings where these two papers were originally presented, has pointed out that enabling developing countries to 'leapfrog' in the adoption of new energy efficient technologies could accomplish this goal.

The need for local studies of causal relations in systemic combinations of populations, consumption, and production is clear, but these local studies need to aim for a general theory that will account for the great variety of local experience. We will illustrate this further below with material from micro-level field research in the Lacandón rainforest.

11.7 What the Population Numbers Mean at the Micro Level: The Lacandón Rainforest

We illustrate several things with the example of the Lacandón rainforest. First, we stress the complexity of the relationship between population and the environment and the number of social, political, historical, and economic factors, *in particular combinations*, that affect their interplay. Second, we introduce the views of the people themselves. Attitudes about the environment and childbearing and their value must be incorporated into models and programs aimed at sustainability.

The colonization and deforestation of the Lacandón rainforest in southeast Mexico began at the turn of the century with the extraction of precious woods by foreign companies until the 1960s and by national government agencies until 1988. Government-sponsored colonization of the rainforest began as part of the Alliance for Progress agricultural settlement programs in the 1960s; it slowed down in the 1970s but faced with the influx of Guatemalan guerrillas and refugees between 1982 and 1988, migration into the rainforest increased along the border with Guatemala. Voluntary migration to the rainforest also occurred as Indians came down from the highlands in the 1970s, pushed both by demographic pressure on lands and by land concentration in the hands of the politically powerful families left intact even after the Mexican Revolution. In 1988 the government of President Carlos Salinas de Gortari decreed a total ban on the cutting of trees in the Lacandón rainforest, which is still being enforced. Programs to develop a sustainable agriculture and agroforestry in the region have been reinforced.

Our research focused on perceptions of deforestation and was based on a survey of 432 households both in the Palenque area and the Marqués de Comillas area deep in the rainforest area at the corner of the border with Guatemala.

Table 11.4 Average number of children per woman and survival rate by community

Communities	Average	Rate
Pico de Oro/Ref. Agraria. Old *ejidos*	5.8	88
Victoria/Nvo. Chihuahua. Poorest *ejidos*	6.9	85
Lacandón/La Unión. Deforested *ejidos*	5.5	83
Palenque Bajos. Low-income group	4.2	90
Palenque Altos. High-income group	2.6	90
Cattle ranchers	2.6	92
Average	5.1	87

Source Arizpe/Velazquez/Paz (1996): *Survey in Lacandon Rainforest*

Table 11.5 Average number of children per woman and survival rate by ethnic identity

Identity	Average	Rate
Indians	6.5	76
Non-Indians	4.8	87

Source Survey in Lacandón rainforest by Arizpe/Paz/Velazquez (1996)

Average number of children are displayed in Table 11.4 as well as the significant range of variation in these averages between areas, communities, and by relative wealth.

The population-environment debate as described in the preceding sections would lead us to ask: do the low-income groups have more children because they are poor or are they poor because they have more children? The question is, of course, simplistic and its answer would be misleading. In fact, the data from the fieldwork and from the survey show how much more complex this issue is.

For example, Table 11.4 contains puzzling data on the survival rates of offspring, which do not support a simple correlation as would be predicted by demographic transition theory. Some of the patterns do seem to be predicted by the theory. The highest survival rate is indeed that of urban Palenque children, due to medical services and higher incomes; the high-income women, knowing their children will survive, have lowered the average number of children they have to 2.6. The lower-income group in Palenque, despite having a high survival rate, still has twice the number of children of the affluent group. Perhaps this is because many of them are migrants who brought with them rural fertility rates or because in unemployment situations having more children is useful.

On this second possibility, when asked what the greatest threat is in the world today, the low-income Palenque group had the highest rate of concern about poverty, even higher than the shifting cultivators of the rainforest. Yet, in spite of this perception, they have a lower average number of children. Another slight discrepancy is that the group with the lowest survival rate of children does not have the highest average number of offspring. Other factors, clearly, are at work. We will consider only two here: ethnicity and religion.

Table 11.5 breaks down our sample according to ethnic identity. A startling difference in the survival rate between Indian and non-Indian children is evident, and it appears to correlate with a very high average number of offspring per woman among Indians. Interviews found that family planning has made no inroads among Indian families, for any number of reasons, including lack of access to family planning programs, resistance by men, isolation, and young marriage age.

That the influence of such factors can be overridden is demonstrated, however, by data on other ethnic groups in Mexico who have lowered their fertility rate in the past two decades (Dirección General de Estadística 1991). This has involved cultural change, women's education and autonomy, and a lessening of geographical and economic isolation.

Similarly, religion also has major effects on fertility levels. Catholic women in the survey have an average of 5.1 children and those belonging to non-Catholic sects, including Jehovah's Witnesses, Seventh-Day Adventists, and Evangelists, average 5.9, in contrast to 2.9 for nonbelievers. But such figures do not fit neatly into patterns. Although child survival is highest among non-Catholic sects it has not led to a lower number of children. However, once again, data show that the religious factor can be overridden, since Catholic women in the Palenque region— with access to schools, clinics, and mass media—have an average of 5.3 children as compared to those in the Marques de Comillas area, who have an average of 6.6 children.

In our view, for any given group it is the *combination* of factors—geographical accessibility, schools, income, women's range of activities, ethnicity, religion, access to information, mass media—that is more important than any single factor. We propose an *inverted nested model* for understanding this complexity. While the nested approach is well suited to studies starting from the global-national level and moving to the local level, an inverted nested model could be useful when the research begins at the local level. It would allow identification of the most salient factors affecting population reproduction at the local level, but would help explain why certain factors take on more salience in particular settings. Thus, for example, the same combination of factors leads to a slightly different average number of offspring between Catholic women in Palenque and in Marqués de Comillas because of the relative importance of different factors.

To the question 'what is the greatest threat in the world today?' only 3.2 % of the total sample referred to overpopulation. Surprisingly, most of them were men and most were cattle ranchers of Palenque. They see the greatest threat to their cattle-ranching business as invasions by landless peasants and are concerned that these peasants should not continue to reproduce at the present rate. A nested approach would allow us to identify population expansion as a concern of both shifting cultivators and cattle ranchers; because of the combination of other factors, these groups end up assigning a different priority to overpopulation as a problem.

The question is no longer why poorer people have more children. Instead it has to be rephrased: 'what is the nested combination of the most relevant factors in a given local situation that leads to given fertility levels?' This implies a much more time-consuming and qualitative approach to data, but the explanatory potential, we believe, is much greater.

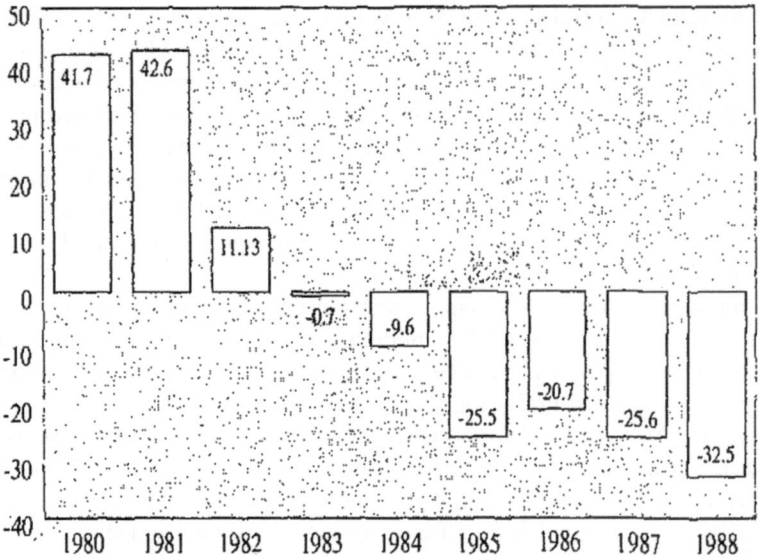

Fig. 11.3 Reversing resource flows: north to south net transfers (US $ billions). *Source* United Nations Development Programme 1990

11.8 Population and Human Development

Notable progress in human development made during this century and especially in recent decades makes it even more difficult to accept our present predicament. On a world scale according to the World Resources Institute (1990: 256), life expectancy has risen from 54.9 (1950–1970) to 61.5 (1985–1990). In developing countries, average infant mortality decreased from nearly 200 deaths per 1,000 live births to about 80 in about four decades (1950–1988), "a feat that took the industrial countries nearly a century to accomplish" (UNDP 1990: 2). Primary health care was extended to 61 % of the population, and safe drinking water to 55 % and despite the addition of 2 billion people in developing countries, the rise in food production exceeded the rise in population by about 20 % (UNDP 1990: 2).

In spite of this progress, in 1985 more than a billion people in developing countries were trapped in absolute poverty, with some groups living in poverty also in developed nations. In 12 of the 23 developing countries where such a comparison is available, the income of the richest groups was 15 times or more that of the poorest group, notably in Latin America (UNDP 1990: 22–23). FAO estimates that about 30 million agricultural households have no land and about 138 million are almost landless, two-thirds of them in Asia (UNDP 1990). A major conclusion of research is that some 500 million to 1 billion poor rural women in developing countries suffer the greatest deprivation, "For them, there has been little progress over the past 30 years" (UNDP 1990: 33).

Importantly, inequities in the distribution of financial and human capital did not decrease but actually grew in the 1980s, both within and among nations. This is illustrated in Fig. 11.3, which shows the financial flows from developing to developed countries. What recommendations have been given to solve this reversal in the fragile progress in human development? The authors of UNDP's 1990 Report on Human Development conclude that "resumed economic growth is thus essential to allow the expansion of incomes, employment and government spending needed for human development in the long run. Without some end to the continuing debt and foreign exchange crisis in much of Africa and Latin America, the impressive human achievements recorded so far may soon be lost" (UNDP 1990: 36).

Clearly, the range of variables impinging on the population question is much more diverse than is usually taken account of in the ongoing population debate in the United States and Europe.

11.9 Population and Resource Use in Social Contexts

The gradual decline in human mortality since the end of the last century must be considered one of the greatest achievements of Western civilization, both on scientific and on human management grounds. Rarely in history has such a unanimous, concerted action been successfully undertaken by such a large number of human agents: scientists, doctors, firms producing medicines, voluntary groups, pharmaceutical companies, and governments—even if some may be guided also by their own economic or political interests.

The Hippocratic oath defending human lives over and above any other consideration gave philosophical sustenance to their activities. Many other cultures, though, do not give such preeminence to the struggle against death; for some it means coining closer to spiritual liberation, or an opportunity for the soul to transmigrate into other realms or beings. Importantly for our discussion, many cultures, especially those living in inclement natural habitats, did subordinate individual human life to the survival of the group. A vast array of fertility control and abortive practices were, and in some cases still are, present in many non-Western cultures. The point is that for centuries many societies had evolved some kind of accounting system whereby the number of people in their group and their age structure were thought of in relation to available natural resources. This was especially true in hunting and gathering, pastoralist and horticultural societies, less so in agrarian societies, where food cultivation could be expanded and more people fed. Other, more aggressive societies obtained the resources needed to sustain their populations through warfare.

These built-in social and ecological accounting systems in many indigenous cultures placed the responsibility for managing sociodemographic processes on the societies themselves. This has been greatly undermined as a result of four driving forces. First, the centralization of power led to the subordination of rural societies to the needs of the urban systems. Second, the loss of cultural diversity eroded cohesive social mechanisms as traditional societies were pulled into market economies and subjected to uniform educational and media systems. Third, due to the spread of urban culture, people are no longer in direct contact with the natural sources of the things they eat, use, or play with, and so lose their bearings as to the depletion of natural stocks. Indeed, the steel and concrete urban environment gives the impression that goods appear purely out of thin air through technological manipulation. Urbanites, however, are becoming keenly aware of the piling up of human bodies in cities, thereby reinforcing their view that the problem is the number of human bodies occupying space and competing for goods, rather than the overall relationship of the urban pattern of consumption of natural resources and planetary stocks. Fourth, scientific models have tended to leave out the cultural and social matrices in which population processes are embedded, and thus have undermined local and meso-level capabilities of organizing their own sociodemographic processes.

Such socially operated cultural accounting systems, which foster local and regional social management, must be revitalized in societies around the world. In developing regions this would allow communities and local peoples to adjust their reproductive behaviour to real expectations of sustainable livelihoods, natural resource availability, and locally defined measures of quality of life. Adapting to the environmental limits to growth in developing countries, however, need not entail accepting the economic limits imposed by continuing subsidies for the wasteful, polluting, affluent life-style of some sectors in industrialized countries of the North.

11.10 Curbing the Growth in Population and in Resource Use

A global perspective of the population-resource use issue means that a reduction in population growth in underconsuming nations must go hand in hand with reducing consumption among affluent groups and nations. This can only be achieved by lowering birth and death rates, alleviating poverty, reducing pressures on resources, and improving women's opportunities, employment-generating policies, and health care (Repetto 1986). At present the debate in the North deals more with population policies to implement in the South rather than with curbing overconsumption in the North (Worldwatch Institute 1988).

Population control is known to be insufficient: it has repeatedly been shown that it is not easily achieved in and of itself and that important social and economic transformations, such as reduction of poverty, must accompany it. "Population can only be expected to fall when livelihoods are secure, for only then does it become rational for poor people to limit family size" (Chambers 1988; Sen/Grown 1988). According to a World Bank study of 64 countries, when the income of the poor rises by 1 %, general fertility rates drop by 3 % (Lappe/Schurman 1988). Reduction of poverty may be a necessary condition for decreasing fertility; it is not, however, a sufficient condition, as the cases of Kerala, Sri Lanka, and other regions demonstrate (Gordon/Suzuki 1991). Nor do lower population growth rates translate immediately to improved environmental conditions. Even in those cases where population growth has been successfully controlled, as in China, the welfare of the people has not necessarily improved and the environment is not necessarily exposed to lower rates of hazard.

To reduce pressures on resources, research priorities should look at situations where demand, either subsistence or commercial, becomes large relative to the maximum sustainable yield of the resource, where the regenerative capacity of the resource is relatively low, or where the incentives and restraints facing *the* exploiters of the resource are such as to induce them to value present gains much more highly than future gains (Repetto/Holmes 1983). Natural resource scarcity studies indicate that a transition will have *to* be made during the next century from cheap, plentiful use of oil to inherently less desirable sources of energy (Mackellar/ Vining 1987), although some authors are more optimistic about unlimited availability of energy (Gilland 1986).

As to the problem of food, prudent optimism largely prevails as to the possibilities of increasing agricultural productivity to feed the increase in population through the year 2000 (Srinivasan 1987; Mackellar/Vining 1987). Some authors, however, are not as optimistic (Brown 1983). To analyze such possibilities, the real problem of production of more food must be separated from the economic and political problem of hunger, that is, of food distribution—the food versus feed issue. Biotechnology provides grounds for optimism, although it seems that its commercial applications will not be seen immediately.

Deforestation, on the other hand, presents a rather more pessimistic picture although different sources cannot agree on the rates of deforestation (Mackellar/ Vining 1987; FAO 1990; Williams 1991). In 1950, industrialized countries imported 4.2 million square meters of tropical woods; in 1980 they imported 66 million (Myers 1981). The outcome will depend on whether consumption of tropical woods and population pressures on the fringes of tropical rainforests are decreased.

11.11 Toward a Global Society

It would seem contradictory to argue in this chapter that one of the driving forces behind the population 'bomb' was that population, as a variable, was abstracted from actual societies with highly disparate natural resource and income distribution bases, and yet emphasize that a 'global society', another abstract construction, must be built. Indeed, we do agree with the Brundtland Commission that there will be no future if we are unable to build one world. To be more precise, we would say that a global society must begin to be interpreted as such, so that it can be seen as such, and therefore built (Arizpe 1991).

But the answer is that such a global society must be built in the same way that the nation-states have been built. Almost without exception, they are internally plural in ethnic and religious identities, per capita income, economic regionalisms, demographic growth rates, and so on, yet juridically and politically they function as a unit. In other words, almost without exception, the unity of nation-states is not an empirical reality, yet the transactions of national and international life are undertaken on the basis of this unity.

In the same way, one can posit that a global society has become a juridical, political, and even cultural necessity, yet the global empirical reality will always be made up of nations and societies, themselves made up of a plurality of trends, some converging, others diverging, that are still not fully understood or susceptible to being totally controlled. They can, however, through negotiations, be successfully managed and pointed in the right direction—if a direction can be agreed upon. Thus, abstracting population as a single factor in models purporting to represent complex empirical reality is inappropriate, but dealing with population as one of the main issues in building a global society is not only appropriate but necessary.

The deeper issue here is one that underlies debates all the way from the Lacandón rainforest in Mexico to the United Nations General Assembly: who is going to build this new economic and accounting system for the world? This is, indeed, apolitical issue at an international level. Since nations are still trying to enhance their own 'wealth of nations', never having left the harbour of classical economics, each will try to build a system that, minimally, will keep its own interests untouched or, maximally, will increase its benefits.

At a more local level, the question of who is creating the new rules, of a global society is perceived in more immediate terms as who is going to bear the cost, actual or potential, of preventing or adapting to new conditions. Whether the debate engages rainforest cattle ranchers and indigenous peoples on deforestation or poor urban dwellers and rich urbanites on urban pollution or corporations and ecologists on economic development or the North and the South on the future of the world, what is at stake is the capacity of human beings to negotiate a common future. And for this purpose, the concept of humanity seems more germane than that of population.

Acknowledgments   Many of the ideas expressed in this section were discussed at meetings of the SSRC/ISSC/DAWN project described in the preface. Our thanks to Richard Rockwell, Gita Sen, William Clark, Rosina Wiltshire and Alberto Palloni.

References

Alba, Francisco; Potter, Joseph, 1986 "Population and development in Mexico since 1940: An interpretation", in: *Population and Development Review*, 12,1 (March): 47–75.

Arizpe, Lourdes, 1978: *Migración, etnicismo y cambio económico* (Mexico City: El Colegio de México).

Arizpe, Lourdes, 1982: "Relay Migration and the Survival of the Peasant Household", in: Balán, Jorge (Ed.): *Why People Move* (Paris: UNESCO): 187–210.

Arizpe, Lourdes, 1991: "The Global Cube: Microsocial Models in a Global Context", in: *International Social Science Journal: Global Environmental Change*, XLIII,130: 599–609.

Arizpe, Lourdes; Costanza, Robert; Lutz, Wolfgang, 1992: "Population and Natural Resource Use", in: Dooge, J.C.I.; Goodman, G.T.; la Rivière, J.W.M. (Eds.): *An Agenda for Science in Environment and Development* (London: Cambridge University Press).

Asian Development Bank, 1991: *Population Pressure and National Resource Management: Key Issues and Possible Actions* (Manila: Asian Development Bank).

Blaikie, Piers M; Brookfield, Harold, 1987: *Land Degradation and Society* (London: Methuen).

Blaxter, Kenneth Lyon, 1986: *People, Food and Resources* (Cambridge: Cambridge University Press).

Bongaarts, John; Mauldin, W. Parker; Phillips, James, 1990: *The Demographic Impact of Family Planning Programs* (New York: Population Council Research Division).

Brinley, Thomas, 1961: *International Migration and Economic Development* (Paris: UNESCO).

Brown Harrison, 1954: *The Challenge of Man's Future* (New York: The Viking Press).

Brown, Lester, 1983: "Global Food Prospects: Shadow of Malthus", in: Glassner, Martin Ira (Ed.): *Global Resources: Challenges of Interdependence* (New York: Praeger).

Caldwell, John C., 1984: *Desertification: Demographic Evidence 1973–1983* (Canberra, Australia: Australian National University, Development Studies Centre).

Chambers, Robert, 1988: *Sustainable Livelihoods, Environment and Development: Putting Poor People First* (Brighton: Institute of Development Studies, University of Sussex).

Clark, Colin, 1958: "Population Growth and living standards", in: Agarwala, Amar Narain; Singh, Sampat Pal (Eds.): *The Economics of Underdevelopment* (London: Oxford University Press): 32–53.

Demeny, Paul, 1988: "Demography and the limits of growth", in: *Population and Development Review Supplement*, 14 (winter): 213–244.

Demeny, Paul, 1990: "Population", in: Turner II, B.L.; Clark, William C.; Kates, Robert W.; Richards, John F.; Mathews, Jessica T; Meyer, William B. (Eds.): *The Earth Transformed by Human Action: Global and Regional Changes in the Biosphere over the past 300 years* (New York: Cambridge University Press and Clark University): 41–54.

Dirección General de Estadística, 1991: *Population Census 1990* (Mexico City: DGE).

Douglas, Ian, 1991: "Human Settlements", Paper for the Workshop on Global Change, Office for Interdisciplinary Earth Studies, Snowmass, Colorado, July 28–August 10.

Durning Alan, 1991: "Asking how much is enough", in: Brown, Lester R.; Durning, Alan; Flavin, Christopher; French, Hilary; Jacobson, Jodi; Lenssen, Nicholas; Lowe, Marcia; Postel, Sandra; Renner, Michael; Ryan, John; Starke, Linda; Young, John: *State of the World 1991. A Worldwatch Institute Report on Progress Towards Sustainable Development* (New York: WW Norton and Co.): 153–169.

Eckholm, Erik P., 1982: *Down to Earth: Environmental and Human Needs* (New York: Norton).

Ehrlich, Paul R.; Ehrlich, Anne H., 1991: *The Population Explosion* (New York: Touchstone, Simon & Schuster Inc).

Ehrlich, Paul; Daily, Gretchen C.; Ehrlich, Anne H.; Matson, Pamela; Vitousek, Peter, 1989: "Global change and carrying capacity: Implications for life on Earth", in: DeFries, Ruth; Malone, Thomas (Eds.): Global Change and Our Common Future: Papers From a Forum (Washington DC: National Academy Press): 19–27.

FAO (Food and Agriculture Organization), 1990: *Vital World Statistics* (Rome: FAO).

Gallopín, Gilberto C., 1990: *Global impoverishment, sustainable development and the environment* (Canada: Ecological Analysis Group).

García, Rolando, 1990: "Interdisciplinariedad y sistemas complejos", in: Leff, Enrique (cood.): *Las ciencias sociales y la formación ambiental a nivel universitario* (Mexico City: UNAM).

Gilliand, Bernard, 1983: "Considerations of world population and food supply", in: *Population and Development Review*, 9,2: 203–211.

Gilliand, Bernard, 1986: "On Resources and Economic Development", in: *Population and Development Review*, 12,2: 295–305.

Goldemberg, José, 1993: Personal Comment, Río de Janeiro, Brazil, June.

Gordon, Anita; Suzuki, David, 1991: *It's a Matter of Survival* (Cambridge: Harvard University Press).

Grant, Lindsey; Tanton, John H., 1981: "Immigration and the American Conscience", in: Nash, Hugh (Ed.): *Progress as if Survival Mattered: A Handbook for a Conserver Society* (San Francisco: Friends of the Earth).

Hardoy, Jorge; Satterthwaite, David, 1991: "Environmental problems of the Third World cities: A global issue ignored?", in: *Public Administration and Development*, 11: 341–361.

Harrison, Paul, 1990: "Too much life on Earth?", in: *New Scientist*, 126, 1717 (May): 28–29.

Hem, Warren M., 1990: "Why Are There So Many of Us? Description and Diagnosis of a Planetary Ecopathological Process", in: *Population and Environment: A Journal of Interdisciplinary Studies*, 12,1 (Fall): 1–27.

Hirschman, Albert Otto, 1958: *The Strategy of Economic Development* (New Haven, Connecticut: Yale University Press).

Jacobson, Jodi L., 1987: *Planning the Global Family*, Worldwatch Paper 80 (Washington, DC: Worldwatch Institute).

Johnson, D. Gale; Lee, Ronald D. (Eds.), 1987: *Population Growth and Economic Development: Issues and Evidence* (Madison: University of Wisconsin Press).

Kasun, Jacqueline P., 1988: *The War Against Population: The Economics and Ideology of World Population Control* (San Francisco: Ignatius Press).

Kelley, Allen, 1986: "Review of the National Research Council Report Population Growth and Economic Development: Policy Questions", in: *Population and Development Review*, 12,3 (September): 563–567.

Keyfitz, Nathan, 1991a: *Need We Have Confusion on Population and Environment?* (Laxenburg, Austria: International Institute for Applied Systems Analysis).

Keyfitz, Nathan, 1991b: *From Malthus to Sustainable Growth* (Laxenburg, Austria: International Institute for Applied Systems Analysis).

Kolsrud, Gretchen; Boyle Torrey, Barbara, 1993: "The importance of Population Growth in Future Commercial Energy Consumption", in: White, James (Ed.): *Global Climate Control* (New York: Plenum Press): 127–141.

Lappe, Frances Moore; Schurman, Rachel, 1988: *Taking Population Seriously* (London: Earthscan).

Leff, Enrique, 1990: "Población y medio ambiente. Es urgente detener la degradación ambiental", in: *DEMOS. Carta demográfica sobre México*, 3: 25–26.

Little, Peter D.; Horowitz, Michael M.; Nyerges, A. Endre (Eds.), 1987: *Lands at Risk in the Third World: Local-Level Perspectives* (Boulder, CO: Westview Press).

Lutz, Wolfgang; Prinz, Christopher, 1991: *Scenarios for the World Population in the Next Generation: Excessive Growth or Extreme Aging* (WP-91-22, Laxenburg, Austria: International Institute for Applied Systems Analysis).

Mackellar, F. Landis; Vining, Daniel R., Jr., 1987: "Natural Resource Scarcity", in: Johnson, D. Gale; Lee, Ronald D. (Eds.): *Population Growth and Economic Development* (Madison: University of Wisconsin Press).

Maihold, Gunter; Urquidi, Víctor (comps.), 1990: *Diálogo con nuestro futuro común. Perspectivas latinoamericanas del Informe Brundtland* (Mexico City: Fundación Frederich Ebert).

Meadows, Donella, 1988: "Quality of Life", in: de Blij, Harm J. (Ed.): *Earth 88: Changing Geographic Perspectives* (Washington DC: National Geographic Society): 332–349.

Myers, Norman, 1987: *Not Far Afield: US Interests and the Global Environment* (Washington DC: World Resource Institute).

Myers, Norman, 1981: "Deforestation in the tropics: who gains, who loses?", in: Sutlive, Vinson H.; Altshuler, Nathan; Zamora, Mario D. (Eds.): *Where have all the Flowers Gone? Deforestation in the Third World* (Williamsburg, Virginia: College of William and Mary): 1–21.

OECD (Organisation for Economic Co-operation and Development), 1991: *The State of the Environment* (Paris: OECD).

Repetto, Robert; Holmes, Thomas, 1983: "The role of population in resource depletion in developing countries", in: *Population and Development Review*, 9,4 (December): 609–632.

Repetto, Roberto, 1987: *Population, Resources, Environment: An Uncertain Future* (Washington DC: Population Reference Bureau).

Revelle, Roger, 1976: "The resources available for agriculture", in: *Scientific American*, 235,3 (September): 165–178.

Robert, Costanza (Ed.), 1991: *Ecological Economics: The Science and Management of Sustainability* (New York: Columbia University Press).

Sage, Colin; Redclift, M., 1991: "Population and Income Change: Their role as Driving Forces of Land-Use Change", Paper for the Workshop on Global Change, Office for Interdisciplinary Earth Studies, Snowmass, Colorado, July 28–August 10.

Sánchez, Vicente; Castillejos, Margarita; Rojas Bracho, Leonora, 1989: *Población, recursos y medio ambiente en México* (Mexico City: Fundación Universo Veintiuno AC).

Sen, Gita; Grown, Caren, 1988: *Development, Crises and Alternative Visions: Third World Women's Perspectives* (New Dehli: DAWN).

Simon, Julian, 1990: *Population Matters: People, Resource, Environment and Immigration* (New Brunswick: Transaction Publishers).

Srinivasan, Thirukodikaval Nilakanta, 1987: "Population and Food", in: Johnson, D. Gale; Lee, Ronald D. (Eds.), 1987: *Population Growth and Economic Development: Issues and Evidence* (Madison: University of Wisconsin Press): 3–26.

SSRC/ISSC/DAWN, 1991: "Recasting the Population-Environment Debate: A Proposal for a Research Program" (mimeo).

Toledo, Víctor Manuel, 1990: "Modernidad y ecología. La nueva crisis planetaria", Mexico City, April (mimeo).

UN (United Nations), 1974: *Human Settlements: the Environmental Challenge* (London: Macmillan).

UN (United Nations), 1990: *Global Outlook 2000: An Economic, Social and Environmental Perspective* (New York: UN Publications).

UNDESA (United Nations Department of International Economic and Social Affairs), 1989: *World Population Prospects 1988* (New York: UN Publications).

UNPF (United Nations Population Fund), 1991: *The State of the World Population 1991* (New York: Oxford University Press).

UNDP (United Nations Development Programme), 1990: *Human Development Report 1990* (New York: Oxford University Press).

Whitmore, Thomas M.; Turner II, B.L.; Johnson, Douglas L.; Kates, Robert W.; Gottschang, Thomas R., 1990: "Long-term population change" in: Turner II, B.L.; Clark, William C.; Kates, Robert W.; Richards, John F.; Mathews, Jessica T; Meyer, William B. (eds.): *The Earth Transformed by Human Action: Global and Regional Changes in the Biosphere over the past 300 years* (New York: Cambridge University Press and Clark University): 25–39.

Williams, Michael, 1991: "Forest and Tree Cover", Paper for the Workshop on Global Change, Office for Interdisciplinary Earth Studies, Snowmass, Colorado, July 28–August 10.

World Resources Institute, 1990: *World Resources 1990-91. A Guide to the Global Environment* (New York: Oxford University Press).

Worldwatch Institute, 1988: *State of the World 1988. A Worldwatch Institute Report on Progress Toward a Sustainable Society* (New York: Worldwatch Institute).

Worldwatch Institute, 1990: *State of the World 1990. A Worldwatch Institute Report on Progress Toward a Sustainable Society* (New York: Worldwatch Institute).

Worldwatch Institute, 1991: *State of the World 1991. A Worldwatch Institute Report on Progress Toward a Sustainable Society* (New York: Worldwatch Institute).

Part IV
Sustainable Development

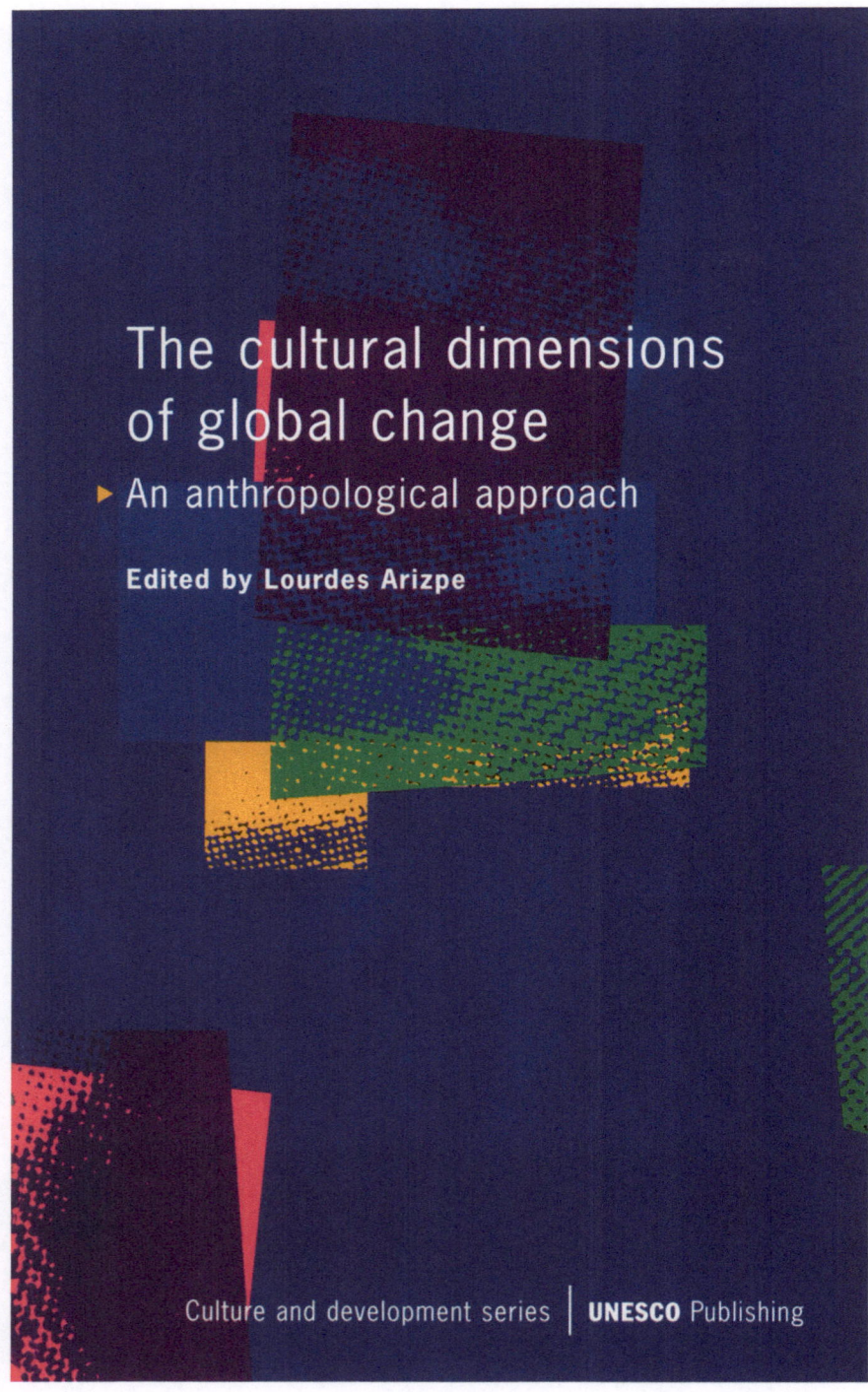

The cultural dimensions of global change

▶ An anthropological approach

Edited by Lourdes Arizpe

Culture and development series | **UNESCO** Publishing

Book cover of Lourdes Arizpe (Ed.), *The cultural dimensions of global change: An anthropological approach* (Paris: UNESCO, 1996)

Chapter 12
Population and Natural Resource Use

Lourdes Arizpe, Robert Costanza and Wolfgang Lutz

12.1 Introduction

Human use of natural resources has continued to increase dramatically in recent years, as a result of complex linkages between increasing population and increasing *per capita* consumption. This has begun to have significant effects on the planet's climate and long-term life-support functions. Future projections of human population range over an order of magnitude by the year 2100, depending on assumptions about rates of fertility and mortality, and indicate a significant aging and urbanizing of the population.

On a finite planet, growth in resource use and population is not sustainable indefinitely. To achieve a sustainable pattern of resource use and population we must understand and control the interactions of population and *per capita* resource consumption as mediated by technology, culture, and values. We must also be careful to differentiate between economic growth (increase in quantity of production and consumption) and economic development (increase of human well-being without a necessary increase in consumption). Technological optimists assume that new technology will be developed to eliminate any resource constraints to continued economic growth. Technological sceptics argue that while this *might* happen it is irrational to count on it. Because of the large uncertainty about the long-term impacts of population and resource use growth on ecological sustainability, we should *at least provisionally assume the worst* and plan accordingly.

This text was originally published as: Lourdes Arizpe, Robert Constanza and Wolfgang Lutz, 1992: "Populations and Natural Resources use", in: Dooge, J.C.I.; Goodman, G.T.; la Rivière, J.W.M. (Eds.): *An Agenda of Science for Environment and Development into the 21st Century* (Cambridge: Cambridge University Press): 61-68. Permission was granted on 28 May 2013 by Ms. Claire Taylor, Senior Publishing Assistant, Legal Services, Cambridge University Press; Cambridge, UK.

We need to focus research on the understanding and control of linked population, consumption, and distribution patterns and not count on a technical fix and continued economic growth to solve the problem.

12.2 Interactions Between Human Populations, Technology, Values and Natural Capital

The crux of the problem of assuring a sustainable world is understanding the full range of possible interactions between, and among, humans and their natural environment, and choosing from this spectrum forms of interaction that are sustainable. While acknowledging that the sustainability concept requires much additional research, we employ the following working definition: *Sustainability* is a relationship between dynamic human cultural/economic systems and larger dynamic, but normally slower-changing ecological systems, in which (a) human life can continue indefinitely, (b) human individuals can flourish, and (c) human cultures can develop but in which effects of human activities remain within bounds, so as not to destroy the diversity, complexity, and function of the ecological life-support system (Costanza 1991).

In this context the way human populations interact with natural capital (renewable and non-renewable natural resources) is critical. We need to develop a much deeper understanding of the relationships between human populations, their technologies, cultures, and values, and the natural capital they depend on for life support if we are to achieve sustainability. Science and technology alone, however, will not achieve this: it is essential that governments and local peoples are mobilized in a major collective effort to ensure the survival of humankind.

12.3 Trends in Human Population and Environmental Transformations

The one common agreement of all human societies is the wish to overcome death; hence the unprecedented alliance of science, government, economic and cultural agents, and communities which effectively brought down mortality all over the globe. Can such agreements be built on fertility, *per capita* resource use, competition for resources and the environment? Such issues touch needs and values which are at the heart of different societies so it is no wonder that "...in the late 1980s, despite the long history of intellectual efforts invested in the population and resources issue, the field appears to be in disarray, with little prospect for the emergence of scholarly consensus" (Demeny 1988: 237).

While scientific research is becoming more specialized and precise, the debate about the global relationship between population and the environment remains polarized between two positions, each of which has a partial scientific understanding of the problem. One position holds that an increasing population is the principal threat to the environment because of the planet's finite resources[1]; the other that human creativity will continue to find technological solutions to expand the planet's carrying capacity. Neither position represents the state of the art of scientific understanding (Johnson/Lee 1987; Repetto 1987; Keyfitz 1991).

These positions, which have moved from the scientific to the public area in popular renditions of the Ehrlich-Simon debate, focus on uni-causal linear explanations and on macro-level analysis using world-wide historical statistics or aggregate simulations. The debate has not taken into account the wealth of data generated at the micro-level, data which provide information on the determinants of fertility and child mortality, and on patterns of production, consumption and distribution. These complex patterns of human relationships overlay, alter and modify the relation of people to land and natural resources. Science must move away from single factor explanations about numbers[2] of people and resource limits and be directed to the analysis of the increase in *per capita* use of renewable and non-renewable resources rather than to numbers of people alone (Demeny 1988: 217; Harrison 1990; Durning 1991).

A scientific framework must be built overcoming the division between contending camps of North and South, rich and poor, since environmental problems are global and regional in causation and in their effects, and they require a concerted effort to arrive at global solutions. In addition, such a division oversimplifies the complex issues that are really a continuation of over-population and under population, overconsumption and under consumption, and sustainable and unsustainable livelihoods. Adopting a different framework will serve as a corrective in itself: much of the present debate in the North is about population policies to implement in the South rather than about curbing overconsumption in the North (Worldwatch Institute 1988).

No simple correlations can be established between population and environmental transformations. In a recent study it was found that the time scales of population variability are asynchronous with given environmental transformations and recovery (Whitmore/Turner II/Johnson/Kates/Gottschang 1990). The authors stress "the need for caution in using population as a simple surrogate for environmental transformation" (Whitmore/Turner II/Johnson/Kates/Gottschang 1990: 37).

A major heuristic problem is the great regional heterogeneity of population growth. Whitmore et al. conclude that "if the experience of past regional population changes and their accompanying environmental transformations has relevance for the future, the projected global scale population 'levelling out' need not diminish the scale and profundity of global environmental change. This is particularly true on the regional

---

[1] A more extreme position likens human population to a cancerous growth bound to kill its hospitable planet.

[2] We refer to population numbers in the sense that the physical presence of human bodies on the surface of the Earth becomes significant because of what these bodies do: they breathe, eat and use the resources of the Earth. A body which does none of these things has no impact on the earth systems.

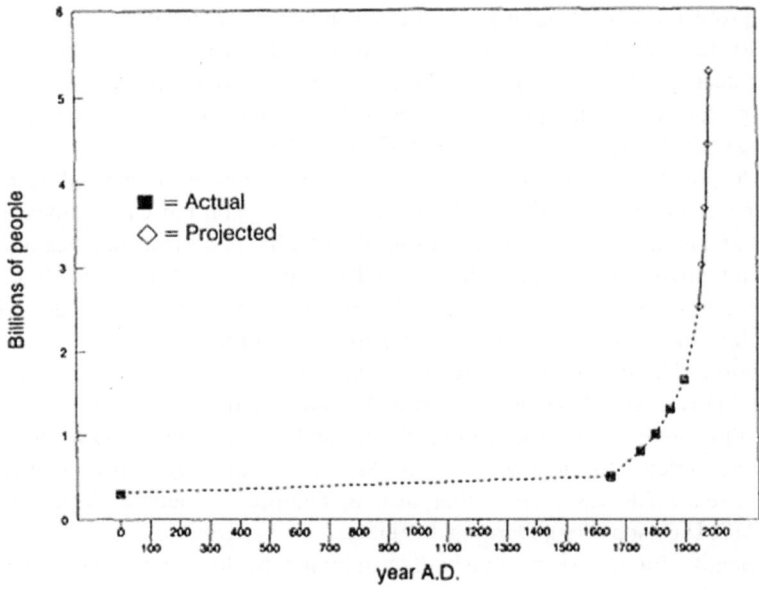

Fig. 12.1 World population from 0 AD

or local scale, where global zero population growth (of population or transformation) need not be accompanied by local or regional equilibrium" (1990: 37).

If one shifts the analysis to the *per capita use* of resources, then priority research areas are the perceptions and assessments of people as to their needs, incomes and desires, as well as their comparative perceptions of the needs, income and desires of other groups. Research on the latter must be extended to analysing the acceptability of different negotiation options as to who is going to bear the costs of adjustments of *per capita* resource use. We need to know much more about the institutional and social capacity for adjustment and adaptation (Jacobson/Price 1990).

12.4 Population Growth up to 1990

With all the qualifications given above, there is little doubt that the recent extremely rapid increase and the projected further increase of the species *Homo sapiens* on our planet is a major factor affecting global resource use which, together with technology and consumption patterns, contributes to the stress put on the natural environment.

Figure 12.1 gives a picture of this extreme acceleration of the growth of the world population. While little is known about prehistorical times, estimates put the total world population of 7000–6000 BC at about 5–10 million people (UN 1973). For millennia there seem to have been fluctuations in population size with a fundamental influence exerted by climatic conditions. Fossil remains show that life was short and frequently ended in violent death. By the year 1 AD the world had an estimated population of 200–400 million. This increase had been made possible by improved agricultural techniques and the development of human civilization. But it was not until the 17th century that population growth started to really take off. The first billion people living on the Earth was reached somewhere around 1830. The second billion was reached in the 1930s; the third in 1960, the fourth in 1975, the fifth in 1987, and the sixth will most likely be reached around 1997. Hence the time it requires to add an additional billion to the world population decreased from more than 100 years for the second billion to about 10 years for the sixth one. Under current growth rates the world population would double again in 40 years.

This unprecedented growth in the number of humans has been brought about by a decrease in death rates while birth rates remained at their traditional high levels.[3] In Europe and North America significant mortality declines were already occurring during the second half of the 19th century, mostly due to improved sanitary conditions and food security. Since then life expectancy has steadily increased in the more developed world. On a global scale the most significant change in mortality came in today's less developed countries (LDC) after World War II with a generally very successful fight against infectious diseases. More recently rates of mortality improvement have declined because the easy measures had already been taken and those requiring infrastructural or behavioural changes proved to be much more difficult.

On the fertility side, birth rates entered a secular decline in Europe and North America during the first decades of this century. This led to a reduction in the growth rate of the population. The phenomenon that birth rates follow the decline of death rates with some lag has been described by the notion of demographic transition. While in the more developed regions this transition is essentially completed today, in most LDCs it is still under way; in some regions—especially Africa—the fertility transition has hardly started.

12.5 Future Population Growth

Changes in population size and age structure have great inertia. Unless wars, famines, or epidemics kill significant proportions of the population, or massive migratory streams empty some regions and fill others, the future population of a certain region can be projected with high certainty in the short run. Projections for the next 20–30 years are rather reliable and insensitive to minor changes in mortality, migration, and fertility. In the following scenarios, we estimate that even widely differing assumptions will yield almost identical results up to the year 2010–2020. But thereafter the range of possible futures opens widely (Lutz/Prinz 1991).

---

[3] In some instances birth rates even increased slightly due to better maternal health conditions.

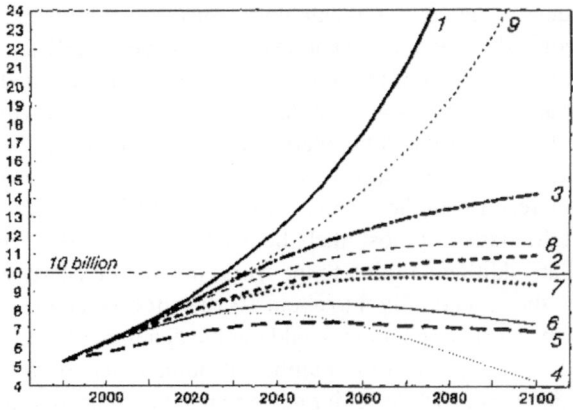

Scenario 1: Constant Rates; constant 1985-1990 fertility and mortality rates
Scenario 2: UN Medium Variant; strong fertility and mortality decline until 2025, then constant
Scenario 3: Slow Fertility Decline; UN fertility decline 25 years delayed, UN medium mortality
Scenario 4: Rapid Fertility Decline; TFR = 1.4 all over the world in 2025, UN medium mortality
Scenario 5: Immediate Replacement Fertility; assumed TFR = 2.1 in 1990, UN medium mortality
Scenario 6: Constant Mortality; TFR = 2.1 all over the world in 2025, constant mortality
Scenario 7: Slow Mortality Decline; UN mortality decline 25 years delayed, TFR = 2.1 in 2025
Scenario 8: Rapid Mortality Decline; life expectancy of 80/85 years and TFR = 2.1 in 2025
Scenario 9: Third World Crisis; constant fertility and 10% increase in mortality in Africa and
Southern Asia; TFR = 2.1 in 2025 and UN mortality for the rest of the world

Note: 'TFR' is the Total Fertility Rate (= average number of children per woman)

Fig. 12.2 Total projected world population 1990–2100 according to scenario

The inertia of population changes also implies that there can be a great impact in the long run of only minor differences in assumed fertility trends over the next few decades. To take an extreme example from the scenarios: whether a certain low fertility level in Africa will be reached by 2025 or 2050 will make a difference in total population size by the year 2100 of more than 1.5 billion, which is almost three times Africa's present population size. This incredible inertia that makes it such a difficult and long-term issue to stop population growth is also called the momentum of population growth, the fact that the age structure of a fast growing population is so young that even if fertility per woman declined to a very low level, the increasing number of young women entering reproductive age will cause the population to grow further for quite some time.

The scenario approach chosen here calculates the implications of several alternative possible future paths of fertility and mortality that need not necessarily reflect the present 'mainstream' thinking. The main value of such a set of alternative scenario projections (that are based on controversial but informed guesses about the future) lies in giving a picture of the possible range of future population sizes and structures. This will help to distinguish almost inevitable trends from changes that are very sensitive to slight modifications in the assumption.

Figure 12.2 plots the total population sizes resulting from the nine scenarios considered here with the projections performed separately for six major regions of the world. In 1990, our planet accommodated 5.3 billion people. Under all scenarios considered over the next 30–40 years, the world population will increase to a size of at least 8 billion. Even immediate replacement fertility in all parts of the world would result in an additional two billion or more people, due only to the momentum of population growth. Under the unlikely rapid fertility decline scenario, assuming only two-thirds of replacement fertility by 2025, the total population size would peak in 2040 at around 8 billion and only decline thereafter.

We may conclude that, on the lower side of the spectrum, unless unexpected major threats to life kill great proportions of the world population over the coming 30–40 years, the world will have to accommodate an extra number of people that is at least as large as half of the world population today. Under the assumption of sustained sub-replacement fertility in all regions of the world, the population might then decline again in the very long run, and possibly by the year 2100 reach a size that is lower than that of today. But still in the transition period the 8 billion mark will be touched.

At the higher end of the spectrum, we have to distinguish between the scenarios that look at the case of continued growth and those assuming a levelling off. Obviously exponential growth cannot continue forever and therefore in the longer run is not only unrealistic but also impossible. Nevertheless, it is instructive to study the results especially in the short- to medium-term future, and compare them to other scenarios. Furthermore, an assumed continuation of currently observed levels is in almost every scientific discipline a standard for comparison unless there is certainty that this level will change in one specific direction. In the case of population growth, we have good reason to assume a change in rates, but we are far from any certainty about the extent and the timing of this decline. In any case, because of the inertia, over the coming three decades the Constant Rates Scenario (1) will not result in very different total population sizes than most other scenarios. Around 2015 the 8 billion mark would be reached and 10 billion only after 2025. Under this scenario a world population of 15 billion would first appear after 2050. For the second half of the next century continued exponential growth would lead to ever-increasing absolute increments resulting, under Scenario 1, in about 40 billion in the year 2100. Further continuation of this growth would then soon result in a 'standing room only' situation. By definition, these exponential scenarios do not assume a feedback from population size to fertility or mortality.

For the scenarios assuming a decline of fertility to replacement level at some point in the future, it appears that even relatively small differences in assumptions concerning the timing of fertility decline have a major impact on population size. Projecting UN Medium Variant assumptions up to the year 2100 gives a levelling off in population growth at around 11 billion, the population size in 2050 already being 10 billion (Scenario 2). Delaying the fertility decline by 25 years, Scenario 3 gives a population size of more than 14 billion in 2100 and population growth does not seem to stop before having reached 15–16 billion in the 22nd century. Likewise, a rapid linear fertility decline to a TFR of only 1.4 children per women in the year 2025 for every region (Scenario 4) gives a totally different picture: After an increase to 8 billion in 2030–2040, population size may decline to a figure below 5 billion in the very long run (2100) as long as no major wars, famines or epidemics occur.

Population size is, to a lesser extent, influenced by assumptions about mortality. In the medium term, a constant mortality level in conjunction with replacement fertility by 2025, Scenario 6, would delay the growth in population size by some 10 years as compared to the UN Medium assumptions on mortality improvements. In the very long run, population size in the absence of mortality improvements tends to level off at around 7.5 billion, which is two billion below that in a corresponding scenario with increasing life expectancy. Delaying the assumed improvement in life expectancy by 25 years, Scenario 7 has virtually no impact on total population size. Assuming a rapid increase in life expectancy to 85 years for women and 80 years for men in all the regions in the year 2025, Scenario 9 increases population size by one billion in 2050 and two billion in 2100.

Understanding the Mesoamerican barter tradition: two women bartering peanuts for avocados in Zacualpan, 2009. Photo is from the author's photo collection

12.6 Age Distributions

The age structure of the population is significant not only because of the above mentioned momentum of population growth, but because it has implications for economic structure and dependency burdens ranging from educational expenses for the young to health care and other support for the elderly. A society's age distribution also has significant impact on consumption patterns, changes in the value system and even culture.

Aging of the population is a universal trend even in less developed countries. While in the more developed countries (MDCs) the proportion of the population above age 60 increased from 11.4 % in 1950 to 17.1 % in 1990 (with some countries such as Germany already above 20 %), in the LDCs this proportion also increased slightly from 6.3 % in 1950 to 6.9 % in 1990 despite the large numbers of new-borns added at the base of the age pyramid.

In the future, further aging seems to be inevitable. Even the Constant Rates Scenario, which is very unlikely in the long run, gives an increase in the mean ages of all regions over the next 50 years. Scenario 2, based on the UN Medium Variant, will result in much more significant aging in all regions of the world. While in today's industrialized countries the mean age is expected to increase from the present 35 years to more than 43 years by 2070, the extent and pace of aging will be even stronger in Eastern Asia. There, the UN Medium Variant expects an increase in the mean age from the current 29 to 44 years by 2070. Even in Africa the mean age is expected to increase by more than 12 years to about the same level we find in Europe and North America today.

As could be expected, the Rapid Fertility Decline Scenario (4), which is the only one that would ultimately bring down the world population below its present size, results in the most extreme aging of the world population. Under this scenario, in almost every continent the mean age of the population would reach about 50 years by the end of next century. What this means in terms of changes of the social and economic structure is hard to imagine, not to mention medical expenses and retirement benefits.

In summary, this comparison of the consequences of various scenarios on total population size and the age structure of the population makes clear the fundamental dilemma of future population trends under low mortality conditions. All scenarios that limit population growth even at a level two to three times of today's world population will result in extreme aging of the population. Only further exponential growth of the population will keep the populations young.

Put in simple words: either the population explodes in size or it ages to an unprecedented extent. The explosion will sooner or later result in higher mortality levels because it cannot go on forever, the aging makes painful social adjustment processes necessary and a complete remodelling of both family and state support systems for the elderly.

12.7 Regional Distributions

The distribution of the population of the major regions of the world has changed significantly over the past years and is likely to change even more in the future. As Table 12.1 indicates, in 1950 one-third of the world population lived in the developed world. In 2000 this fraction will have declined to about one-fifth despite the fact that the population of the developed countries grew by about 50 %.

By the end of our century the population of Africa will have quadrupled from 224 million to 872 million and increased its share of the world population from 9 to 14 %. On the losing end we find Europe with a halving of its proportion from

Table 12.1 Population sizes and distributions, 1950 and 2000

	1950		2000	
World	2,516	(100 %)	6,122	(100 %)
Developed countries	832	(33 %)	1,277	(21 %)
Developing countries	1,684	(67 %)	4,846	(79 %)
Africa	224	(9 %)	872	(14 %)
Latin America	165	(7 %)	546	(9 %)
North America	166	(7 %)	297	(4 %)
Asia	1,376	(55 %)	3,549	(58 %)
China	555	(22 %)	1,256	(21 %)
India	358	(14 %)	964	(16 %)
Europe	392	(16 %)	512	(8 %)
Oceania	13	(5 %)	30	(5 %)
USSR	180	(7 %)	315	(5 %)

The estimates given for 2000 result from Scenario 2 as described

16 to 8 % and North America with a decline from 7 % in 1950 to only 4 % in 2000. Latin America and Asia will have slightly increased their share. Due to its successful population stabilization policy, the share of the Chinese population in the world has even declined slightly.

For the year 2050 the above described scenario calculations yield the following pattern. The proportion of the world population living in Africa will inevitably further increase to as much as 25 % under Scenarios 1 and 9. Even under the rapid fertility decline scenario the proportion will further increase to 17 %. At the other extreme the proportion of the world population living in Europe, North America, and the USSR together would decline to as little as 7 % by the year 2050 under the constant rate scenario. Even under the unlikely rapid fertility decline scenario the proportion in those more developed regions would decline from the present 20 % to around 12 %. This also demonstrates a strong momentum in the growth and shrinking of relative population sizes in the major world regions.

12.8 Migration and Urbanization

The human capacity for adaptation (the highest among natural species) allowed *Homo habilis* to roam the Earth for 3 million years, dispersing all the way from East Africa to Tierra del Fuego in the tip of South America. But never has the magnitude of such geographical shifts been as high as those caused by the demise of agrarian societies since the 17th century. The four major agricultural crises in Europe—mostly Central and Baltic—between 1844 and 1913 sent out most of the 52 million emigrants that went overseas to settle in thinly populated areas (Brinley 1961: 11).[4] With the spread of the market economy in the second half of this

---

[4] Other kinds of outflows have involved fewer numbers: the slave trade across the Atlantic, six to twelve million; political refugees in all countries of the world, 14 million.

century, the rural exodus has become pervasive, though with different outcomes, in Latin America, Africa, and Southern Asia.[5]

In the last few decades, rural outmigration in the latter three regions has changed: push factors have become stronger than pull factors; internal migrations now overflow into the international arena; gender imbalances in migratory cohorts tend to decrease as economic opportunity costs and cultural determinants begin to change the gender division of labour and of geographical mobility, though, at the same time poverty tends to push more women and older people out of rural villages into the cities; and, most recently, outmigration from cities towards other locations is also on the rise.

Projections tend to indicate that migrations will continue to increase in all developing regions, fostered by a combination of factors including the spiral of population growth and poverty; land or wealth concentration; economic polarization in agricultural production; and inefficient, corrupt or mistaken government policies. All of these lead to rural outmigration: loss of livelihoods, land degradation and desertification, and food and land scarcity.[6]

Another important factor, which will very likely increase the importance of the 'pull' factors in rural outmigration, is the spread of a cultural urban bias through education and the mass media (Swaminathan 1988). Additionally, if competition for control and access of scarce resources (land, capital, technology, water) increases (especially in rural areas of less developed countries and in Eastern European countries, most probably running along traditional lines of ethnicity and nationalism) political refugees will also most likely increase in numbers.

Migration has been a survival strategy for the poor, be they a Punjabi family having several sons working in the Gulf region, or a Peruvian couple who live off the remittances sent by their daughter from the United States. Regional

---

[5] "Perhaps nine-tenths of the population increase in Northern America and Oceania and two-thirds of that in Latin America could be directly attributed to European migrant populations within Europe's demographic outshoots in the vast, formerly thinly populated land of the Americas, Oceania and Northern Asia. Altogether the areas of European settlements that comprised 20 % of the world's population in 1700 claimed 36 % of that total by the middle of the 20th century" (Demeny 1990).

[6] A typical combination of such factors at a regional scale in Latin America is given by Stonich 1989): "Agricultural development in the region has been highly uneven not only in terms of the spatial distribution of people. Political-economic factors related to the expansion of export-oriented agriculture constrain access to the most fertile lands of the region (in South Honduras). This has resulted in a highly uneven distribution of population in which the greatest population densities occur in the highlands, the areas that are the most marginal for agriculture. The growing population in the highlands with inadequate opportunities to earn a living has led to a parceling of land among more and more people, with agricultural production expanding into even more marginal areas. Growing rural poverty stimulated outmigration from the more densely patched South into other parts of the country, thereby decreasing population pressure in the region awhile simultaneously augmenting urban populations and escalating pressure on tropical forest areas in the remainder of the country".

studies which analyse the relationship between economic and migration flows are necessary in order to map out the intricate web of migratory movements that may have global effects.

Future research priorities related to migration might seem rather simple; people will continue to flow to where wealth, meaning possibilities of livelihoods, amenities and the picture of the 'good life' are concentrated. Studies that disaggregate migrant cohorts by gender, age, ethnic group, etc. may provide valuable information to 'colour' the maps of migration flows, but will yield few possibilities for prediction. In contrast, studies on the selectivity of migrants, that is who migrates within the local community and *why* they settle at a given destination may provide useful insights for policy and planning options.

At the global and regional level, migration studies should focus on probabilistic and simulation models of major outflows which may be expected with increasing climatic events and with cumulative environmental changes that may destroy local people's livelihoods. First in line would be outflows in Africa due to famines and desertification; and from regions most vulnerable to possible climatic changes, especially coastlines, deltas and islands. Most migration will tend to flow from South to North, although also from East to West and from South to South. A great uncertainty in attempts to forecast international migratory streams are also short term changes in the immigration policies of the MDCs and the question how effectively such policies can be enforced.

Presently 45 % of the world population lives in urban areas (37 % in LDCs, 73 % in MDCs). This percentage is likely to increase to 51 % by 2000 and 65 % by 2025 (UNPF 1991). While urbanization at different levels is a universal trend in less developed regions the rapid growth of some mega-cities is most visible. In 1960 seven of the world's ten largest urban agglomerations were in North America, Japan and Europe, with New York, Tokyo and London at the top. Presently Mexico City is the largest one with more than 20 million, and seven of the top ten are in LDCs.

12.9 Population Pressure and Per Capita Resource Use

A primary question is: are there limits to the carrying capacity of the Earth System for human populations? Kenneth Blaxter gives an unequivocal yes, but cautions that "where doubt sets in is on the precise definition of the number of people that can be supported, about the way in which population will increase, about the way in which food production will reach the limit imposed by the carrying capacity and about the availability of resources to push back this limit" (Blaxter 1986). These issues must be the priority research topics for the next decades. Various estimates of global carrying capacity of the Earth for people have appeared in the literature, ranging from 7.5 billion (Gilliand 1983 cited in Demeny 1988) to 12 billion (Clark 1958), 40 billion (Revelle 1976) and 50 billion (Brown 1954). However, many authors are sceptical about the criteria—amount of food, or kilocalories—used as a basis for these estimates. "For humans, a physical definition of needs may be irrelevant. Human needs and aspirations are culturally determined: they can and do grow so as to encompass an increasing amount of 'goods', well beyond what is necessary for mere survival" (Demeny 1988: 213–244).

Cultural evolution has a profound effect on human impacts on the environment. By changing the learned behaviour of humans and incorporating tools and artefacts, it allows individual human resource requirements and their impacts on their resident ecosystems to vary over several orders of magnitude. Thus it does not make sense to talk about the 'carrying capacity' of humans in the same way as the 'carrying capacity' of other species (Sanchez/Castillejos/Rojas 1989; Blaikie/ Brookfield 1987) since, in terms of their carrying capacity, humans are many subspecies. Each subspecies would have to be culturally defined to determine levels of resource use and carrying capacity. For example, the global carrying capacity for *Homo americanus* would be much lower than the carrying capacity for *Homo indus,* because each American consumes much more than each Indian does. And the speed of cultural adaptation makes thinking of species (which are inherently slow changing) misleading anyway. *Homo americanus* could change its resource consumption patterns drastically in only a few years, while *Homo sapiens* remains relatively unchanged. We think it best to follow the lead of Daly (1977) and Daly/Cobb (1989) in this and speak of the product of population and *per capita* resource use as the *total impact* of the human population. It is this total impact that the Earth has a capacity to carry, and it is up to us to decide how to divide it between numbers of people and *per capita* resource use. This complicates population policy enormously, since one cannot simply state a maximum *population*, but rather must state a maximum number of *impact units.* How many impact units the Earth can sustain and how to distribute these impact units over the population is a very dicey problem indeed, but one that must be the focus of research in this area.

Many case studies indicate that "there is no linear relation between growing population and density, and such pressures (towards land degradation and desertification)" (Caldwell 1984). In fact, one study found that land degradation can occur under rising pressure of population on resources (PPR) under declining PPR and without PPR (Blaikie/Brookfield 1987). Therefore, the scientific agenda must look towards more complex, systemic models where the effect of population pressures can be analysed in its relationships with other factors (García 1990). This would allow us to differentiate population as a 'proximate' cause of environmental degradation from the concatenation of effects of population with other factors as the 'ultimate' cause of such degradation (Asian Development Bank 1991).

Research can begin by exploring methods for more precisely estimating the total impact of population. For example, Clark (1991) suggests that the 'Ehrlich identity' (Pollution/Area = People/Area × Economic Production/ People × Pollution/Economic Production) can be operationalized as (CO_2 Emissions/km^2 = Population/km^2 × GNP/Population × CO_2 Emissions/GNP). Clark and his colleagues examined data for twelve countries from 1925 to 1985 and concluded that the same loading of pollution on the environment can come from radically different combinations of population size, consumption, and production. Thus no single factor dominates the changing patterns of total impact across time. This points to the need for local studies of causal relations among specific combinations of populations, consumption and production, noting that these local studies need to aim for a general theory that will account for the great variety of local experience (see also Lutz 1991).

Along this line of research, Torrey and Kolsrud (1991, 1993) examined population growth and energy efficiency in several countries and concluded that the very small population growth forecast for developed countries over the next forty years will add a burden of CO_2 emissions that will be equal to that added by the much larger population growth forecast for the less developed countries. Decreasing energy consumption in developed countries could dramatically decrease CO_2 emissions globally. It is only under a scenario of severe constraints on emissions in the developed countries that population growth in less developed ones plays a major global role in emissions growth. If energy efficiency could be improved in the latter as well as the former, then population increase would play a much smaller role. Goldemberg (1993) has suggested that enabling developing countries to 'leapfrog' in adopting new energy efficient technologies could accomplish this goal.

Research priority should also look at situations where demand (either subsistence or commercial) becomes large relative to the maximum sustainable yield of the resource, or where the regenerative capacity of the resource is relatively low, or where the incentives and restraints facing the exploiters of the resource are such as to induce them to value present gains much more highly than future gains (Repetto/Holmes 1983).

12.10 Curbing Population Growth and Per Capita Resource Use

Some authors single out a high rate of population growth as a primary root cause of environmental degradation and overload of the planet's carrying capacity, and therefore point to population control as the appropriate policy instrument. Ehrlich and his colleagues maintain "There is no time to be lost in moving toward population shrinkage as rapidly as is humanly possible" (Ehrlich/Daily/Ehrlich/Matson/Vitousek 1989). Certainly, one step we should take immediately is to provide universal access to simple birth control measures to allow potential parents the full range of options in planning their families. Concerning population policies the same standards, such as voluntary choice of family size and family planning methods, should be applied in both more and less developed countries. Policies focusing solely on the control of population size are known to be insufficient. It has repeatedly been shown that the level of fertility in a population cannot be simply regulated but depends crucially on social and economic changes such as improved educational status, especially of women, improvements in social infrastructure and the reduction of poverty. There are also examples showing that curbing population growth does not necessarily result in an improvement of welfare and of environmental conditions, although one could make the argument that conditions might be even worse if rapid population growth had continued.

The opposite, position is taken by those who see high rates of population growth as stimulating economic development through inducing technological and organizational changes (Hirschman 1958; Boserup 1965), or as a phenomenon which can be solved through technological change (Simon 1990). Aside from the important question whether there is enough time for such reactions to be effective, such positions ignore the dangers of environmental depletion implicit in unchecked economic growth: consumption increases and rapidly growing populations that can put a very real burden upon the resources of the Earth, and bring about social and political strife for control of such resources. This position also assumes that technological creativity will have the same outcomes in the future as in the past, and in the South as in the North, a questionable assumption. Finally, it heavily discounts the importance of the loss of biodiversity—a loss which is irreversible and whose human consequences are as yet unknown.

A different approach is taken by those authors for whom, "population can only be expected to fall when livelihoods (of the poor) are secure, for only then does it become rational for poor people to limit family size" (Chambers 1988; Sen/Grown 1988). According to a World Bank study of 64 countries, when the income of the poor rises by 1 %, general fertility rates drop by 3 % (Lappe/Schumann 1988). When making such statements, however, one must be aware of the great social and cultural heterogeneity among the poor in different parts of the world which is highly relevant for the way in which their fertility reacts lo improving living conditions. In contrast to the previous positions, some authors state that 'population is not a relevant variable' in terms of resource depletion and stress that *resource consumption,* particularly overconsumption by the affluent, is the key factor (Hardoy/Satterthwaite 1991; Sánchez 1990; Harrison 1990; Durning 1991). OECD countries represent only 16 % of the world's population and 24 % of land areas; but their economies account for about 72 % of the world gross product, 78 % of road vehicles, 50 % of global energy use. They generate about 76 % of world trade, 73 % of chemical products exports, and 73 % of forest product imports (OECD 1991). The main policy instrument, in this ease, in the short term is then, reducing consumption and this can be most easily achieved in those areas where consumption *per capita* is highest.

Thus a new framework should expand the definitions of issues: focus not only on population size, density, rate of increase, age distribution, sex ratios, but also on access to resources, livelihoods, social dimensions of gender, and structures of power. New models have to be explored in which population control is not simply a question of family planning but of economic, ecological, social and political planning (UNDP 1990; Jacobson 1987); in which the wasteful use of resources is not simply a question of finding new substitutes but of reshaping affluent lifestyles and in which sustainability is seen not only as a global aggregate process but also as one having to do with sustainable livelihoods for a majority of local peoples.

Acknowledgment   Much of this chapter is taken from the previous work of the authors, which benefited from the collaboration of many individuals, among them: Herman Daly, Joy Bartholomew, Richard Rockwell, Charles Perrings, Cutler Cleveland, John Cumberland, Margarita Velázquez, Bruce Hannon, Bob Ulanowicz, Christine Halvorson, Veronica Behn, and Alan Scholefield.

References

Asian Development Bank, 1991: *Population Pressure and National Resource Management: Key Issues and Possible Actions* (Manila: Asian Development Bank).

Blaikie, Piers M; Brookfield, Harold, 1987: *Land Degradation and Society* (London: Methuen).

Blaxter, Kenneth Lyon, 1986: *People, Food and Resources* (Cambridge: Cambridge University Press).

Boserup, Ester, 1965: *The Conditions of Agricultural Growth: The Economics of Agrarian Change under Population Pressure* (Chicago: Aldine).

Brinley, Thomas, 1961: *International Migration and Economic Development* (Paris: UNESCO).

Brown, Harrison, 1954: *The Challenge of Man's Future* (New York: The Viking Press).

Caldwell, John C., 1984: *Desertification: Demographic Evidence 1973–1983* (Canberra, Australia: Australian National University, Development Studies Centre).

Chambers, Robert, 1988: *Sustainable Livelihoods, Environment and Development: Putting Poor People First* (Brighton: Institute of Development Studies, University of Sussex).

Clark, Colin, 1958: "Population Growth and Living Standards", in: Agarwala, Amar Narain; Singh, Sampat Pal (Eds.): *The Economics of Underdevelopment* (London: Oxford University Press): 32–53.

Clark, William C., 1991: "Environment and Risk", Paper presented at the Annual Meeting of the American Association for the Advancement of Science, Washington DC.

Costanza, Robert (Ed.), 1991: *Ecological Economics: The Science and Management of Sustainability* (New York: Columbia University Press).

Daly, Herman E., 1977: *Steady State Economics* (San Francisco: W. H. Freeman).

Daly, Herman E.; Cobb, John B. Jr., 1989: *For the Common Good: Redirecting the Economy toward Community, the Environment and a Sustainable Future* (Boston: Beacon).

Demeny, Paul, 1988: "Demography and the Limits of Growth", in: *Population and Development Review Supplement*, 14 (winter): 213–244.

Demeny, Paul, 1990: "Population", in: Turner II, B.L.; Clark, William C.; Kates, Robert W.; Richards, John F., Mathews, Jessica T.; Meyer, William B. (Eds.): *The Earth Transformed by Human Action: Global and Regional Changes in the Biosphere over the past 300 years* (New York: Cambridge University Press and Clark University): 41–54.

Durning, Alan, 1991: "Asking How Much is Enough", in: Brown, Lester R.; Durning, Alan; Flavin, Christopher; French, Hilary; Jacobson, Jodi; Lenssen, Nicholas; Lowe, Marcia; Postel, Sandra; Renner, Michael; Ryan, John; Starke, Linda; Young, John: *State of the World 1991. A Worldwatch Institute Report on Progress Towards Sustainable Development* (New York: WW Norton and Co.): 153–169.

Ehrlich, Paul; Daily, Gretchen C.; Ehrlich, Anne H.; Matson, Pamela; Vitousek, Peter, 1989: "Global change and carrying capacity: Implications for life on Earth", in: DeFries, Ruth; Malone, Thomas (Eds.): *Global Change and Our Common Future: Papers From a Forum* (Washington DC: National Academy Press): 19–27.

García, Rolando, 1990: *Metodología para el estudio del cambio global* (Mexico City: ISSC Workshop Issues in Global Environmental Change).

Gilliand, Bernard, 1983: "Considerations of World Population and Food Supply", in: *Population and Development Review*, 9,2: 203–211.

Goldemberg, José, 1993: Personal Comment, Río de Janeiro, Brazil, June.

Hardoy, Jorge; Satterthwaite, David, 1991: "Environmental problems of the Third World cities: A Global Issue Ignored?", in: *Public Administration and Development*, 11: 341–361.

Harrison, Paul, 1990: "Too Much Life on Earth?", in: *New Scientist*, 126,1717 (May): 28–29.

Hirschman, Albert Otto, 1958: *The Strategy of Economic Development* (New Haven, Connecticut: Yale University Press).

Jacobson, Jodi L., 1987: *Planning the Global Family*, Worldwatch Paper 80 (Washington, DC: Worldwatch Institute).

Jacobson, Harold K.; Price, Martin F., 1990: *A Framework for Research in the Social Sciences for Global Environmental Change* (Paris: ISSC).

Johnson, D. Gale; Lee, Ronald D. (Eds.), 1987: *Population Growth and Economic Development: Issues and Evidence* (Madison: University of Wisconsin Press).

Keyfitz, Nathan, 1991: "Population and Development within the Ecosphere One View of the Literature", in: *Population Index*, 57,1(spring): 5–22.

Kolsrud, Gretchen; Boyle Torrey, Barbara, 1993: "The importance of Population Growth in Future Commercial Energy Consumption", in: White, James (Ed.): *Global Climate Control* (New York: Plenum Press): 127–141.

Lappe, Frances Moore; Schurman, Rachel, 1988: *Taking Population Seriously* (London: Earthscan).

Lutz, Wolfgang, 1991: "Towards the Holistic Understanding of a Micro Cosmos a Case Study on Mauritius", in: *Population, Economy and Environment in Mauritius* (CP-91-01, Laxenburg, Austria: International Institute for Applied Systems Analysis): 3–8.

Lutz, Wolfgang; Prinz, Christopher, 1991: *Scenarios for the World Population in the Next Generation: Excessive Growth or Extreme Aging.* WP-91-22 (Laxenburg, Austria: International Institute for Applied Systems Analysis).

OECD (Organisation for Economic Co-operation and Development), 1991: *The State of the Environment* (Paris: OECD).

Repetto, Roberto, 1987: *Population, Resources, Environment: An Uncertain Future* (Washington DC: Population Reference Bureau).

Repetto, Robert; Holmes, Thomas, 1983: "The Role of Population in Resource Depletion in Developing Countries", in: *Population and Development Review*, 9,4 (December): 609–632.

Revelle, Roger, 1976: "The resources Available for Agriculture", in: *Scientific American*, 235,3 (September): 165–178.

Sánchez, Vicente, 1990: *Población y medio ambiente* (Mexico City: Fundación Medio Mundo).

Sánchez, Vicente; Castillejos, Margarita; Rojas Bracho, Leonora, 1989: *Población, recursos y medio ambiente en México* (Mexico City: Fundación Universo Veintiuno AC).

Sen, Gita; Grown, Caren, 1988: *Development, Crises and Alternative Visions: Third World Women's Perspectives* (New Dehli: DAWN).

Simon, Julian, 1990: *Population Matters: People, Resource, Environment and Immigration* (New Brunswick: Transaction Publishers).

Stonich, Susan C., 1989: "The Dynamics of Social Processes and Environmental Destruction: A Central American Case Study", in: *Population and Development Review*, 15,2 (June): 269–296.

Swaminathan, Mankombu Sambasivan, 1988: "Global agriculture at the crossroads", in: de Blij, Harm J. (Ed.): *Earth 88: Changing Geographic Perspectives* (Washington DC: National Geographic Society): 316–331.

Torrey, Barbara; Kolsrud, Gretchen, 1991: "Population Growth and Energy Consumption", Paper presented at the Annual Meeting of the American Association for the Advancement of Science, Washington DC.

UN (United Nations), 1973: *The Determinants and Consequences of Population Trends* (New York: United Nations).

UNDP (United Nations Development Programme), 1990: *Human Development Index* (New York: UNDP).

UNPF (United Nations Population Fund), 1991: *The State of the World Population 1991* (New York: Oxford University Press).

Whitmore, Thomas M.; Turner II, B.L.; Johnson, Douglas L.; Kates, Robert W.; Gottschang, Thomas R., 1990: "Long-Term Population Change" in: Turner II, B.L.; Clark, William C.; Kates, Robert W.; Richards, John F., Mathews, Jessica T.; Meyer, William B. (Eds.): *The Earth Transformed by Human Action: Global and Regional Changes in the Biosphere Over the Past 300 years* (New York: Cambridge University Press and Clark University): 25–39.

Worldwatch Institute, 1988: *State of the World 1988. A Worldwatch Institute Report on Progress Toward a Sustainable Society* (New York: WW Norton and Co).

Cultura y Cambio Global: Percepciones Sociales sobre la Desforestación en la Selva Lacandona

Lourdes Arizpe
Fernanda Paz
Margarita Velázquez

Centro Regional de Investigaciones Multidisciplinarias
Universidad Nacional Autónoma de México
Grupo Editorial Miguel Ángel Porrúa

Book cover of Lourdes Arizpe, Fernanda Paz and Margarita Velázquez *Culture and Global Change: Social Perception of Deforestation in the Lacandona Rainforest*, first published by CRIM-UNAM and M.A. Porrúa, 1995. Also published in English by the University of Michigan at Ann Arbor,1996. This photo was taken by CRIM-UNAM

Chapter 13
Human Dimensions of Global Change

13.1 Introduction

For the past two decades phenomena that appeared as isolated processes, such as deforestation, ozone depletion, the extinction of some animal species, and air pollution, are now being perceived as part of a broader pattern of global change.[1] The challenge for social scientists is to study local environmental and social changes that are associated with global trends. This field of study is being carved out as the human dimensions of global environmental change (Jacobson/Price 1990). Calling it 'global environmental change' points to the fact that there may also be research fields on global economic change or global cultural change, although it has been suggested that the term *globalization* be used mainly to refer to such social processes.

The phenomena of global change are rapidly fostering new research collaboration between natural and social scientists and are certainly becoming some of the major issues for international negotiations in the years to come. As an emerging field, however, there is still much to be done to throw light on the scope and implications of global change. Developing countries should take this unprecedented opportunity to take part in the early stages of a research process that should lead to more balanced and equitable proposals for global development.

The term *global change* is only just beginning to be used and is thus still in the phase of having its 'human dimensions' mapped out. Basically, the concept of 'global' is used to refer to those phenomena that affect all the Earth's inhabitants. It is also being used, however, to designate an emerging new perspective of world

---

[1] This text was originally published as a chapter on: "Human Dimensions of Global Change": in: Arizpe, Lourdes; Paz; Fernanda; Velázquez, Margarita: *Culture and Global Change: Social Perceptions of Deforestation in the Lacandona Rainforest in Mexico* (Ann Arbor: University of Michigan Press, 1996): 9–18. Permission to republish this text in this anthology was granted by Aaron McCulloug and Debra Shafer on 9 July 2013. Printed copies of the original book may still be obtained at: http://www.press.umich.edu/14545/culture_and_global_change.

L. Arizpe, *Migration, Women and Social Development*,
SpringerBriefs on Pioneers in Science and Practice 11,
DOI: 10.1007/978-3-319-06572-4_13, © The Author(s) 2014

events and to label a new stage in human civilization, a new global era. In other words, it is being used to designate both empirical phenomena and a new theoretical field, which might be the starting point for creating a new scientific and political paradigm (Burton/Timmerman 1989). There are even those who would insist that a field of global science be developed.

Our problem in the world today is that we have to describe, analyse, reflect on, and propose solutions and create new institutions all at the same time, a situation that is only comparable to the formative period of industrialism three centuries ago. Yet today our time is running out. As a result, science is not only expected to analyse the causes and impact of global changes but also to suggest effective proposals to direct them with as much certainty as possible. This means there is a need to change the demarcations between the sciences themselves (i.e., the categories of exact, natural, and social sciences) and between them and their 'associates for change,' as the International Council of Scientific Unions (ICSU) has called governments, nongovernment organizations, international organizations, and the private sector.[2]

In this context, then, how should global change be defined? If we limit ourselves to the idea that only phenomena affecting the entire human race are global, it would only include those involved in biogeochemical cycles, such as the greenhouse effect or the depletion of the ozone layer. Its study, by definition, would fall within the area of the physical and natural sciences.

Yet almost 80 % of global environmental changes are caused by human actions, and 100 % of the solutions to these changes also depend on human actions.[3] Therefore, since most biogeochemical phenomena are anthropogenic, the solutions that have to be found are made up of the decisions and actions of individuals and societies. Environmental global change, then, is inseparable from the human dimensions of global change, although the structure of scientific disciplines is biased toward a heuristic division in these different fields of study. The important point, obviously, is that joint models of the physical, natural, and social sciences need to be developed for the study of global phenomena.

But, just as 20 % of global phenomena are exclusively dependent on biogeochemical causes and not anthropogenic ones, many areas of socio-political change associated with globalization are also exclusively dependent on social, economic, political, and cultural factors. Thus, in addition to the basic human well-being that depends on the planet's well-being, there is another area of human well-being that depends mainly on economic, social, and political development.

The field of global studies, then, is made up of two overlapping circles, one of the natural sciences and the other of the social sciences. While they share a concern for global change and globalization, their specific approaches come from one field or the other. The global perspective emerged from the satellite images that

---

[2] In May 1990 the International Council of Scientific Unions (ICSU) held a meeting on 'Science and Its Partners' to establish a dialogue between universities, private sector companies, and international science organizations, and governments on the challenges of global change.

[3] T. Rosswall, personal communication, committee meeting on Human Dimensions of Global Change, April 1990, Paris.

gave us the breath-taking photograph of our blue planet, but it is replayed every day in the instant images we get through telecommunications of events on the other side of the planet.

The dilemma posed by the new globality that we now see in images and know from statistics and the fragmented cosmopolitical scheme of segmentary sciences and of nation-states, tribes and localities, was most accurately captured by the United Nations (UN) Environment and Development Commission, in the first sentence of its report *Our Common Future* (Brundtland Commission 1987): "The Earth is one, but the world is not". There are some issues, then, that are now unequivocably global, the major one being that a new threshold has been crossed whereby the human species might become extinct if the bioatmospheric conditions needed to sustain human lives atrophies. Faced with this new idea, humans, only too humanly, tend to resort to old behaviour patterns, withdrawing into tight-knit fidelities, whose expression might be political, nationalistic, or ethnic. This disjunction is at the core of what social scientists should be studying and politicians organizing in order to bring human behaviour patterns into accord with the new global realities.

13.2 The Scientific Challenge

To put it simply, the two main scientific challenges related to global change from our point of view involve creating models to include both the natural and social aspects of global phenomena and explaining the interactions between the local and the global levels. The exact and natural sciences study the scales, rates, and forms of interaction of phenomena such as emissions of greenhouse gases; climatic change; ozone depletion; air, rain, and water pollution; loss of biodiversity; soil erosion; and changes in land use. The social sciences should thus focus on studying the human actions that cause these changes, identifying those who are responsible and those who would be most vulnerable to the impact of these global environmental changes and the new forms of political negotiations and world institutions that must be developed to solve these environmental problems at all levels and locations. These research areas fall within the broad fields of styles of development and of the social perceptions and assessments that different nations or groups establish in analysing their interests in relation to global environmental change.

To determine which direction to take and to identify priority issues in the social science research of global change, the first challenge involves devising a general scheme, or 'wiring diagram,' such as the one produced by the commission led by Francis Bretherton for physical and natural sciences to study global change. We attempted this exercise at a 1991 workshop on global change led by Bretherton himself, with the participation of fifty natural, physical, and social scientists.[4] The

---

[4] Workshop on Global Change, sponsored by the Office for Interdisciplinary Earth Sciences, in which fifty natural, physical, and social scientists participated, 27 July–10 August, in Snowmass, Colorado.

result, preliminary thus far, is a model that does not represent empirical phenomena but, rather, specifies the information requirements that must be supplied by the different sciences in order to study global change.

The second challenge is to construct models that include biophysical and social perspectives of these phenomena. Several groups of scientists are already working on a project of this sort. Their aim is to create models to establish a typology of the planet's most endangered ecosystems, linking them to human actions that affect them and to the socio-political processes that sustain and foster these actions.

The third challenge regarding the human dimensions of global change consists of analysing the relationship between regional and global phenomena. Two problems arise here. The first is that regional phenomena have a certain autonomy in terms of global dynamics, as shown by studies of historical demography (Turner/Clark/Kates/Richards/Mathews/Meyer 1990). At present, there are no models that accurately represent the relationship between these two levels. The second problem is that biogeophysical systems and social systems do not always coincide in space and time. At present, data that are obtained, especially through geographical information systems, are georeferential (i.e., they establish matrices based on geographical parameters), yet the social processes that drive them may in fact come from other geographical regions. For example, deforestation in the Lacandona forest is the result of the economic dynamics of the country's south eastern region but also of the political decisions made by the government in Mexico City, of the consumer habits of urban populations, and of the demographic and land tenure trends in other states. Thus, the types of data and analytical models that need to be constructed to explain this phenomenon fully go beyond the limits of georeferenced data on the Lacandona region.

The scientific challenge in the study of global change also includes other fundamental aspects. Among other things, it has to do with the interpretative frameworks being created to contextualize these studies at a time when scientific paradigms are changing. Natural and exact sciences are now interested in studying nonlinear, irreversible, and unstable phenomena, research that will draw them closer to the social sciences, whose focus has always included these features.

A further challenge, already mentioned in previous pages, involves constructing new theoretical models to study societies that are experiencing increasingly swift, fluid changes. An additional complication is that such changes must not only be studied but managed as well. This is the socio-political challenge in the field of global change.

13.3 The Socio-Political Challenge

The main risk is that at the same time we are heading toward 'one world,' as proposed by the Brundtland Commission (1987), we are also drifting toward many worlds: some are islands of prosperity and have high protectionist walls; others have oceans of poverty that will become the misery of the new millennium. In this context a global perspective may help focus on the need to achieve a more balanced development for all regions and nations, taking into account cultural and geographical specificities.

In any case, one would have to ensure that the call for a single world would encourage creation of new international agreements and institutions and the gradual development of a global civic culture. What should be avoided is allowing this universal mandate, as the word *universal* was used in the past, to veil the reestablishment of an organization that will favour only the most developed countries.

It would be too long to review the well-known globalization processes associated with current economic and social development. Instead, we will concentrate on the topics presently being suggested as top priorities regarding global environmental change from a social perspective. The International Social Science Council's Program on the Human Dimensions of Global Environmental Change has proposed six key topics in its framework for research in this field[5]:

1. social dimensions of the use of resources: production, reproduction, and consumption
2. perceptions and evaluations of global conditions and environmental changes
3. impact of social, economic, and political structures at local, national, and international levels
4. land use
5. energy production and consumption
6. industrial growth.

These topics specifically concern problems associated with the use of natural resources and global change. Some of the main issues involved are:

1. the changes in the use of raw materials and imported agricultural products in developed economies that can change patterns of natural resource use in developing countries
2. the challenges for future industrialization in southern countries because of problems posed by fossil fuel consumption and the emission of greenhouse gases, requiring the transfer to clean technology and investment resources
3. the relationship between population growth, increased consumption of natural resources, food production, and emission of pollutants, particularly, the growing flow of 'ecological refugees'
4. the possibilities of raising agricultural productivity, taking into account severe problems of soil erosion, pollution linked to the use of fertilizers and pesticides, changes in regional microclimate, and access to agricultural biotechnologies
5. the protection of tropical rain forests with alternatives of productive development for farming communities in these forests

---

[5] The International Social Science Council, Standing Committee on Human Dimensions of Global Environmental Change; see Jacobson and Price (1990). The committee is headed by Harold Jacobson and in its initial phase included Lourdes Arizpe, Daniel Bertaux, Ashish Bose, Takashi Fujii, Leszek Kosinski, Kurt Pawlik, Renat Perelet, Martin Price, and Robert Worcester.

6. the differential vulnerability of various social and ethnic groups to the impact of global environmental phenomena
7. the ways in which environmental changes are perceived and evaluated by different groups involved in political negotiations to solve socioenvironmental problems and their repercussions on democratization processes
8. the cultural values that establish patterns of well-being and consumer habits that influence economic and social development and the role of the mass media.

13.4 The Invisible Sociosphere

The social sciences will have to undergo a transformation to meet the challenge of analysing and explaining global change. Let us begin by pointing out that the 'sociosphere,' the social counterpart of the terms *geosphere* and *biosphere,* cannot be seen in a photograph of the planet taken from outer space.[6] As a construct, it exists only in our minds. Consequently, and for a number of other reasons, this has created the impression that transformations of the geosphere and biosphere are purely natural phenomena, whereas, in fact, most are the result of human use of the planet's resources.

Physical and natural sciences have made great strides in their knowledge of how, where, and how fast the natural environment's resources are being used. Social sciences should concentrate on why these resources are being used in specific ways—in other words, for what purpose, and, crucial to our finding solutions for a sustainable future, who is using them. It is essential that we identify the network of interests that often hinder programs and actions aimed at achieving sustainable development.

While this field of research is beginning to be known as global change, Price (1989) points out that it has mainly been defined from geocentric and biocentric points of view. In other words, there is a need to move to a more balanced perspective that will include a human and social dimension.

To do this, it is not enough simply to prepare a list of research topics in relation to global change. There is a need for a new global perspective, and there are already a number of proposals within the social sciences for creating this broader view. Burton and Timmerman (1989: 302) believe that a "new paradigm" based on what they call "the development of complex systems" should be built. This would not only mean creating a new theoretical research program but also reconsidering some of the epistemological bases of current social sciences. Since these are intrinsically linked to nineteenth-century philosophical

---

[6] Daniel Bertaux discussed this point at one of the meetings of the ISSC UN and UNESCO Standing Committee on Human Dimensions of Global Change. We are grateful to the committee members for the debates, which provoked many of the reflections described in this book.

principles, new philosophical and ethical norms that are relevant to the present time must be explored. Gallopín and researchers at the Bariloche Foundation in Argentina have suggested the concept of 'global impoverishment' to include both ecological and economic impoverishment as a central process of global change (Gallopín/Gutman/Maletta 1989; see also Gutman 1988 and Leff 1986). What criteria can we use in the social sciences, then, to define global phenomena?

13.5   Definition of Global Change in the Social Sciences

Natural scientists have already identified the world's most pressing environmental problems in the *International Geosphere-Biosphere Program: A Study of Global Change* (ICSU 1989). Menon, former president of the International Council of Scientific Unions, has defined global environmental change as follows (1989: 60):

> Humankind has no doubt been altering the environment in the process of living and development for at least 2 million years, but during most of this time, human influence on the environment has been local in scale and small in magnitude. It is only over the last half a century that humankind has developed the ability to alter the environment on a global scale, and not just in terms of local effects such as due to pollution.

From this point of view the criterion for the definition of the global concept is one of *scale*. Additionally, Menon also mentions, in his description of the key objective of the International Geosphere Biosphere Programme (IGBP), that global change involves "interactive physical, chemical, and biological processes that regulate the earth system, the unique environment that it provides for life, the changes that are occurring in this system, and the manner in which they are influenced by human actions" (Menon 1989: 60). Thus, the second criterion used to define it, then, is the *interaction* between the different processes that are seen as its components.

If the main criteria used to define global phenomena in natural sciences are scale and interaction, it would be easy to create a parallel scheme of these phenomena in the social sciences, simply by adding another level of magnitude to our models, that of 'global,' to the already existing phenomena of economic systems, telecommunications networks, and so on.

Significantly, however, the changes regarded as global, as far as human groups are concerned, are not only associated with changes in the speed, density, and scale of interactions but also with modifications in the structure and complexity of these interactions. As Miller (1989: 87) points out, in order to explain world environmental change, it is necessary to examine the direct human actions that produce an effect on it as well as the indirect human actions that set in motion complex sequences of events that also affect the environment. The study of the indirect human actions involved in the dynamic of changing political and economic systems is the reason that the development of specific social theory to explain the global level is necessary. Thus, research into global change should go beyond simply measuring what can be seen; it should tackle the more difficult task of establishing the assumptions and parameters for interpreting this change.

To put it bluntly, since there is no history of global phenomena in human experience, not only do we lack the methods to study them; we also have no basic categories or ideas to help us think about them. Thus, the traditional body of theoretical thought in the social sciences offers very few hypotheses to interpret these phenomena, which helps explain why they are so invisible from the point of view of social science. Only one social science, anthropology, has tried to tackle what is not being conceptualized as globality.

13.6 The Anthropological Experience

Despite the fact that anthropology ended up as the study of 'others'—that is, of the 'peoples without history'—in its basic assumptions it is global in scope and intent. This should be understood in the sense that its field of study has covered all the world's peoples in the present as well as the past and in their interaction with the natural environment.

Information gleaned on this last point has shown, among other things that this is hardly the first time in history that ecological factors have forced civilizations along a given path of development or decline. One example of this would be the Egyptian civilization, whose appearance has been linked to the desertification of neighbouring regions, which drew agricultural populations to the fertile Nile valley and led to the emergence of a highly developed civilization (Manzanilla 1989). There are also many examples of complex states whose collapse was the result of political and social uprisings associated with the depletion of surrounding environmental resources, as may have been the case of the Mayan city of Copan (Abrams/Rue 1988).

A different exercise focusing on attaining a global view of human development was the attempt by many anthropologists to draw up a world map of all cultures. The most important of these was based on comparative data from George Murdock's Human Relations Area Files (1967). There are many similarities between the problems of theory and method put forward in Murdock's project and those currently being examined in relation to the data and models for global environmental change; we shall mention the two most important ones.

The main problem is how to establish the heuristic limits of the units of analysis. In cultural analysis this involves deciding whether a culture should be considered a single unit, despite its internal diversity, or whether its subcultures should be considered as separate entities. This methodological decision may alter cultures' statistical distribution and therefore the generalizations that can be made regarding cultural traits or institutions throughout the world. Thus, these boundaries pose not only technical problems of statistical measurement but also theoretical problems related to the definition of existing relationships between the different cultures. There may be many differentiated cultures belonging to an overarching cultural tradition, as was the case in the Mayan region.

The same kinds of problems arise when the human dimensions of global environmental change are studied. How does one establish a relationship between significant social groups and the georeferenced processes of geoatmospheric models? How can political, administrative, national, state, provincial, and municipal boundaries, for which statistical data exist, be used to study environmental depletion? How can ethnic boundaries be related to environmental phenomena?

Global change, associated both with environmental and human phenomena, essentially constitutes the crossing of a historical threshold to enter a new age in human civilization. This transition is difficult to understand and analyse using the classical instruments of social sciences, so fundamental changes in assumptions and theories must be made. Furthermore, we should look toward the future, both to analyse what is happening and also to contribute to the formulation of a new cosmopolitical scheme, a new ethics, and the new social pacts necessary to ensure the sustainability of a global society.[7]

References

Abrams, Elliot M.; Rue, David J., 1988: "The Causes and Consequences of Deforestation among the Prehistoric Maya", in: *Human Ecology*, 16: 377–395.

Brundtland Commission (UN World Commission on Environment and Development), 1987: *Our Common Future* (New York: Oxford University Press).

Burton, I.; Timmerman, P., 1989: "Human Dimensions of Global Change: a Review of Responsibilities and Opportunities", in: *International Social Science Journal*, 121 (August): 279–313.

Gallopín, G.; Gutman, P.; Maletta, H., 1989: "Global Impoverishment, Sustainable Development and the Environment", A Report to IRDC on the Project 'Global Impoverishment and Sustainable Development (Mimeo).

Gutman, Pablo, 1988: *Desarrollo Rural y Medio Ambiente en América Latina* (Buenos Aires: CEAL).

ICSU (International Council of Scientific Unions), 1989: *The International Geosphere-Biosphere Programme: A Study of Global Change* (Paris: ICSU).

Jacobson, Harold K.; Price, Martin F., 1990: *A Framework for Research on the Human Dimensions of Global Environmental Change* (Paris: HCGEC-ISSC-UNESCO).

Leff, Enrique (Ed.), 1986: *Los Problemas del Conocimiento y la Perspectiva Ambiental del Desarrollo* (México City: Siglo XXI).

Manzanilla, Linda, 1989: *La Formación del Estado en Babilonia* (México City: UNAM).

Menon, M.G.K., 1989: "Opening Address", in: *Global Change*, Report 71.

Miller, Roberta, 1989: "Human Dimensions of Global Environmental Change", in: De Fries, R.S.; Malone, T.C. (Eds.): *Global Change and our Common Future* (Washington: National Academy Press): 84–89.

Murdock, George, 1967: *Ethnographic Atlas* (Pittsburgh: University of Pittsburgh Press).

Price, Martin, 1989: "Global Change: Defining the Ill-defined", in: *Environment*, 31,8 (October): 18–20, 42–44.

Turner, Billie Lee II; Clark, William C.; Kates, Robert W.; Richards, John F.; Mathews, Jessica T.; Meyer, William B., 1990: *The Earth as Transformed by Human Action: Global and Regional Changes in the Biosphere over the Past 300 Years* (New York: Cambridge University Press—Clark University).

---

[7] We wish to acknowledge Benjamin Mayer's participation in the analysis for this section.

Speaking at a meeting on rural development in New Delhi, 1988. This photograph is from the author's photo collection

Chapter 14
Culture and Sustainability

Lourdes Arizpe and Fernanda Paz

14.1 Introduction

The concept of sustainability is similar to that of democracy, difficult to define in different social settings, elusive in its applications, yet essential to mark a landing point in the horizon of the future. Both are an ideal of socio-political behaviour yet neither can be analysed as to actual results until they have been embodied in everyday practices in different societies.

Therefore, there are two kinds of questions we can ask to try to understand what sustainable development means. The first is whether sustainability is an attribute of culture, that is, an ideal which needs to be worked out in practice. The second is whether sustainability is the complex and cumulative result of negotiations between social actors on the ground. For this analysis the assumption is made that development models are the common political goals of a society that bring together such ideals and negotiations.

Where does sustainability fit into a development model? Is it in the cultural values behind abstract ideals? In the perceptions and interests that drive different social and political groups to act in certain ways towards the environment? Or is it in the actual biophysical exchange between societies and the natural world? It would seem that the latter is receiving the most attention today in policy discussions and in the mass media. However, the contention of this chapter is that it is the first two driving forces that actually determine that biophysical exchange with the natural environment. It is essential then to explore the meaning of sustainability in relation to these two driving forces.

This text was first presented as a coauthored paper to the International Forum on Sustainable Development, UNESCO, Paris, 23–25 September 1991 and has not previously been published.

14.2 Sustainability: The Ideal of the New Millennium

The idea of sustainable development, as first enunciated by the United Nations Commission for Development and Environment[1] is an expression of a new political will which points, no more and no less, to the desire of humanity to continue living on Earth in ways similar to those we know today. Our problem today is that we know what no longer works yet we have not had the time to try out new forms of international organization, of governance, and of market exchange which will hopefully work.

This explains some of the scepticism that has emerged in relation to sustainability. What is the scientific status of this concept? It is argued, quite short-sightedly in our view, that the term was coined in discussions of policy and that therefore there is no place for it within scientific discourse. Let us not forget, however, that practically all classical concepts used in the social sciences which today have a scientific status also emerged from political philosophy and social commentary.

Those who know about development processes are also sceptical of the view that sustainability can be achieved in the face of powerful macroeconomic and political interests. This reflects a deeper concern that the slowing down of development in an unequal world will only make this inequality more pronounced. Consequently, less-informed groups have taken the position that sustainability— environmental or ecological concerns—is opposed to development, a view which is fortunately being left behind. However, there is some truth in the statement that the North wants sustainability, while the South wants development. The *status quo* does not benefit the North, but its overconsumption of natural resources is just as threatening to the life of the planet as the social under-consumption of the South, so finding a way out of this dilemma entails a very complex and long-drawn-out negotiating process which should be the focus of international relations in the years to come.

In any case, greater conceptual clarity as to what sustainability means is a necessary ingredient for solving this problem. In this paper the task is to compare the concept of sustainability with other similar concepts used in social and anthropological analysis and then to explore whether beliefs can be 'sustainable', using fieldwork data about the inhabitants of the Lacandon rainforest in south-east Mexico. Then, very summarily, the relationships between different groups in the rainforest are described in a search for the factors influencing the 'sustainability' of their actions towards the rainforest.

---

[1] The World Commission on Environment and Development (1983–1987), chaired by Gro Harlem Brundtland, released the Report *Our Common Future*, also known as the *Brundtland Report*, in October 1987, a document which coined and defined the meaning of the term 'sustainable development'.

14.3 Sustainability and Social Reproduction

The Brundtland Commission in *Our Common Future* (1987: 15) defined sustainable development as "development that meets the needs of the present without compromising the ability of future generations to meet their own needs". The problem with this definition according to Ayres (1991) is that no one knows how to measure welfare in social terms. He proposes a more intricate definition: "Sustainability is a process of change in which the exploitation of resources, the direction of investments, the orientation of technological development and institutional change is in harmony and enhances both the current and the future potential to meet human needs and aspirations". As precise as this definition is, it still lands us with the problem of defining 'in harmony'. It is clear that social scientists—anthropologists, sociologists, psychologists—must be brought in to explore these concepts.

The latter is especially true because it is mainly economists who are working on finding a more operational definition of sustainability. Mäler (1990: 240), for example, defines it as "the economic development in a specified area (region, nation, the globe) is sustainable if the total stock of resources—human capital, physical reproducible capital, environmental resources, exhaustible resources—does not decrease over time". The Latin American and Caribbean Economic Commission (CEPAL 1991: 22) has argued that the biological concept that considers a sustainable activity as one which does not go against certain natural laws must be given a wider significance by bringing in other criteria in the management and use of natural resources such as citizen participation, policy decision-making, and institutions. The emphasis is on the principle that sustainability should not be defined primarily by the amount of natural resources available but by the use made of such resources.

The use of natural resources has to do with the way in which societies reproduce themselves. Interestingly, the concept of sustainability emerged only a few years after the discussion of 'social reproduction' in sociology and women's studies. One use of the concept of social reproduction grew out of an interest within French sociology in trying to explain why certain ideological or political structures were reproduced through the State, or through the actions of the dominant class. Pierre Bourdieu and others also analysed how an authoritarian culture is transmitted through 'symbolic violence' in the school system (Bourdieu/Passeron 1970).

Achieving sustainable development means doing away with beliefs and actions contrary to a harmonious relationship between human life and the geosphere. In this sense it means deterring the reproduction of such beliefs and actions; if sustainability is stated only in terms of goals, of what should be, it will become imprisoned in moral imperatives. Instead, it is necessary to analyse and explain the mechanisms whereby the perceptions and actions deleterious to the environment are reproduced from generation to generation. In this task, the heuristic instruments developed for the analysis of social reproduction may be of use. If, as has been mentioned, domination and inequality play an important role in driving unsustainable practices, then in order to overcome such practices it is important to make the linkages visible between domination and inequality on the one hand and unsustainable practices on the other.

Important linkages can be made with feminist research on the analysis of reproduction in societal terms (Benería/Roldán 1987). Women in all societies are, mythically and empirically, seen as the embodiment of social reproduction, to our great honour, of course. Feminist analysis, then, has built a series of theoretical schemata of reproduction—biological, labour, communal and social—as a complementary and integrated field to that of production, which is socially defined as that of men. One of the common denominators of such schemata is the need to overcome gender and other inequities in order to achieve a harmonious reproduction of human and social relationships. It is, indeed, a quest for sustainability, in this case defined as the minimization of interpersonal and social violence and suffering to ensure a nurturing environment for new generations. The lesson to be taken from this is that sustainability must mean a harmonious relationship not only with the natural world, but with the human world as well. The concept of sustainability must have a dimension of *social* sustainability.

Another important aspect is that up until now the discussion of sustainability has emphasized the macroeconomic relationship between society and the natural environment. But unless we begin to think of sustainable strategies at the level of the local household the result will be oversimplified generalizations having little to do with the actual everyday practices of women and men.

The proposal is that a comparative analysis should be undertaken which looks at the rationality and consistency of sustainable strategies at the macroeconomic level and compares them with those of households at the local level. This is very important and an example will illustrate why. Population growth is one of the crucial driving forces in the use of natural resources in the world today, yet the dysfunctionality of this trend at the macro-environmental level contrasts with the functionality of this strategy at the household level in particular social settings, for example, in certain agrarian communities having a large number of offspring is actually one of the best strategies to cope with encroaching poverty. Also, it is well known that women play a central role in curbing population growth, mostly through their education and employment status which has to do with the gender and age distribution of labour, and with authority within the household.

A final interesting point is the widespread generalization found in many cultures whereby women are considered to be close to nature while men are seen as close to the political and economic organization of society. Notwithstanding the implied discriminatory feature of such representation, since it seems to come from deep cultural layers in societies, it is foreseeable that women will also begin to be linked to sustainable possibilities for the future, just as indigenous societies are already considered to act in more environmentally sustainable ways than their non-indigenous counterparts. This a priori perception, though, must be borne out in empirical reality, where, not surprisingly (but for *social* not *genetically* determined reasons!), this perception is true, although there are instances where it is not.

14.4 Cultural Extinction and the End of History

The will to prevail and to expand geographical and politically seems to be one of the prime movers of human cultures throughout history. In the same way that a life instinct is postulated as an inherent trait in the psychological make-up of human beings, so an instinct to survive and to prevail can be postulated as a social imperative in all human societies. In fact, there has never been a society in the records of historians and anthropologists that has consciously and deliberately moved, as lemmings do, to collective self-destruction. In this respect, the collective suicide (although it seems not all were willing) of more than seven hundred people of a religious sect in Georgetown, Guyana in the seventies can only be construed as an alarm signal for the whole of humanity. *Something is wrong. Social sustainability, just like environmental sustainability, seems to be imperilled.* The one thing we cannot cope with is suicide. Albert Camus was right when he said that the only serious problem in philosophy is suicide. This means that the reverse is true, that an instinct of survival is present in all individuals and all cultures. How much more unfathomable, then, is the problem posed by the possibility of collective suicide by the whole of humanity if the life-supporting basis of the planet is destroyed.

However, history shows that even if this 'instinct' is present, and even if social and political organizations support sustainability, many cultures have disappeared because sustainability may be an effect derived from a *systemic combination* of factors not all of which are susceptible to human control, for example political upheaval linked to environmental factors. Thus a society's endogenous capability for sustained development does not always ensure or at least historically has not always ensured its survival.

Moreover, one of the most intractable problems in historical interpretation has to do with assigning causality to the factors that intervened in the downfall or disappearance of a civilization or culture. For example, in Copan, Honduras, the palynological records indicate that the theocratic state declined at the same time that agricultural fields and food-gathering areas began to be situated at an increasing distance from the ceremonial centre. There is also evidence, though, as in the case of other Maya centres, that there may have been social and political disruptions. Which came first, the depletion of nearby natural resources or political upheaval? In all probability they fed into each other: a political rebellion caused by extreme political demands from the central theocracy or scarcity of food may have caused the breakdown of the food production system which in turn had declined because of the depletion of nearby natural resources.

The point to be made here is that practically all political processes deal intrinsically with the allocation of resources. Therefore, a programme for sustainability is, *ipso facto*, a programme for political change.

For this reason, it would be an attempt to trivialize to state that sustainable development is just another fashionable term in the end-of-the-century political discourse. On the contrary, it reminds us of a profound human will to prevail which seems to be increasingly suffocated by the urban comforts and consumerism of postmodern society in the North, and by desperate misery among the poor in the South.

Furthermore, if we believe that the starting point of history was the rise of human consciousness about infinite natural resources then we are indeed at the end of history. Not the minutely brief history of political confrontation which (Fukuyama 1989, 1992) claims has been won by liberal democracy, but the invention of a bountiful world when peoples all around the world have created an image of themselves with or towards the natural world.

It is important to realize that almost all advanced civilizations developed along fertile valleys and coastlines which gave support to a perception of infinite natural resources to be used. In some cases where such plentiful resources were not present the alternative plan of action was to have infinite human resources (meaning in many cases enslaved societies) that could be organized to bring to the urban centre resources from faraway lands.

In this light the empire of modernity which proclaimed the ideal of infinite progress was standing on an Age of Exploration which said that natural resources were infinite and could be harnessed for human development Western-style. Indeed, this cornerstone of the philosophical edifice of the Enlightenment would have crumbled had the Europeans not been able to navigate through the seas of the world to take control of the extraction of these resources and of agricultural production in order to accumulate capital in Europe. Thus, it can be said that the ideal of infinite progress was underhandedly supported by the political and military capacity of western countries to ensure unlimited accessibility to prime natural resources.

We are now at the end of a period of at least 200 years in which such unlimited control and use of natural resources both by capitalist and socialist industrialism went unquestioned. Now it is being questioned on political as well as on technological environmental grounds.

The realization that we live in a world with finite natural resources (even though some argue, rather untenably, that technology will infinitely expand their use) opens up a new era. And the ideal for this new era will no longer be unlimited progress but sustainable development. But such an ideal, as mentioned earlier, touches on two different issues: how are the stocks of natural resources being used, and who is using them, that is, who is to bear the cost of changing our environmental course?

It is interesting to see that these issues, now being discussed in international forums on a global scale, are also present and in a much more pressing way in micro-social situations. In what follows, an exploratory analysis is made of how local people in the rainforest of south-east Mexico perceive the question of the sustainability of those valuable resources as well as the problem of the comparative use that different groups make of them.

14.5 Problems Caused by the Way the Lacandón Rainforest was Developed

Until the beginning of the sixties the Lacandón rainforest had lost only 6 % of its original 1.4 million hectares. The first to arrive there in the last century were the European logging companies who used the river to transport timber for export to Europe. They continued these extractive activities both of timber and of rubber until the 1960s. They left very little behind them in the region, only a few roads and surnames among the population.

Although it had been inhabited by an autochthonous group of hunters and gatherers, the Lacandón Indians, the colonization of the rainforest began in the sixties, driven both by a government programme and by the spontaneous migration of Indian peasants from the highlands of Chiapas. Beginning in 1963, the Mexican government, as part of the 'Alliance for Progress' policy, began a colonization programme in the region of Palenque through an agrarian reform settlement scheme. It offered lands to farmers from other regions of Mexico where demographic pressure on land was very high.

The import substitution development model implemented by the Mexican government in the fifties and sixties was based on the assumption that natural resources in the country were unlimited and all that was needed was to exploit them for industrial growth and export. The export of standing timber was seen in this light, as one more way to finance national economic growth.

In 1974, in an attempt to regain control of what had become chaotic immigration to and illegal logging in the rainforest in Chiapas, President Luis Echeverría decreed usufruct rights to 614,321 ha of forest to 321 Lacandón Indians who then held the exclusive right to cut down trees for export in the forest.

As the trickle of spontaneous immigration into the rainforest grew into a continuous flow by the end of the seventies, new legislation was introduced which created the Montes Azules Biosphere Reserve, with an area of 321,200 ha. Its boundaries, though, overlapped with those of the previous decree and the ambiguity has led to the Reserve being invaded by local Indian farmers who say they have a right to those lands.

In the beginning of the eighties, the municipality of Marqués de Comillas in the innermost corner of the Lacandón rainforest on the border with Guatemala became a focus of potential military and political conflict as both Guatemalan guerrillas and refugees poured across the border. To stem these flows, in 1983 the Mexican government decided to create a human border by bringing in settlers, giving them possession of 50 ha of land each and providing them with chainsaws. As a result, it is estimated that of the approximately 198,000 ha of rainforest in Marqués de Comillas, 80,000 ha were deforested between 1983 and 1988.

In December 1988 incoming President Carlos Salinas de Gortari and Chiapas Governor Patrocinio González, going against the vested interests, decreed a total ban on the cutting down of trees in the Lacandón rainforest in an attempt to stop deforestation. To help farmers develop more sustainable kinds of agriculture, several government programmes have been initiated in the region. This has stabilized the farming communities but has had little effect among those settlers interested not in farming but rather in following the example of some of their immediate predecessors who 'got in, got rich and got out', by selling everything on their land plots: woods, wild animals, plants and fish.

The main actors in this rainforest drama, then, are the Indian and *mestizo* farmers, get-rich-quick opportunists, cattle ranchers, government officials and the urban population of Palenque.

Can the 'sustainability' of their perceptions about the natural world around them be assessed in the verbal expression of their perceptions?

14.6 Nature, World, Land: Many Names

"Nature, I don't know what that is, I didn't go to school" said a *mestizo* woman who lived in the rainforest and now ekes out a living in Palenque. A most significant answer because local people there are beginning to have the impression that things that have to do with 'ecology', 'the ozone effect' (sic), and 'deforestation' have to do with high-tech studies. Our team of researchers was constantly being asked what these terms meant. But another reason why she answered in that way is that Nature (*Naturaleza*) is not the term used by the peasants to denote the natural world.

The term they use most frequently is that of 'world' (*mundo*) as a *single, unitary entity*, which means they do not divide the world into a human/social and a natural order, even though most of them hold Christian beliefs. They also use other unitary concepts such as 'everything' (*todo*), or 'things' (*cosas*). The only other more restrictive term they use is that of 'land' (*tierra*) which may mean both the Earth and the soil.

A majority of those interviewed, especially Indian or *mestizo* farmers, gave a creationist view of the natural world. "God made everything" one stated bluntly. "There is a maker of all the natural things" said another, who lives deep in the rainforest, on the Lacantun river, "I don't believe that it was a question of chance and nothing else; a God created heaven, earth, water and animals". Some insist on it even though they have heard about evolutionism: "I almost don't believe in evolution" said a farmer from El Naranjo, "we believe that it was God who created everything; we don't believe in the transformation of times. We were educated like this since we were children".

Inevitably, the Bible is the point of reference on the nature of Nature, especially among many who have converted to Protestant sects—Jehovah's Witnesses, Pentecostalists and Evangelicals. Their interpretations, though, are always woven with poetic licence, as Bernabé Álvarez, a Catholic Chol Indian showed:

> In the beginning God made the heaven and the earth; the earth was pure water like an abyss, there was no dry land. The spirit of God passed over the water like a dove and from there he went up to heaven and thought 'it is no good the way it is'. This is what God thought, then he formed the dry land and the river between the land, and this is how trees and animals were born and after that, the man, Adam. There are fruit trees as well. And as the man grew sleepy, God took a rib to create his woman-companion and then there were two.... Then he created the fish in the water, when the water was still clean; he created everything that is animal on dry earth and when he created them, God said: 'that's it, now it is created'.

But whereas there is substantial consensus on the creationist view of Nature, differences among them arise when one asks, what was Nature—the world, things—created for?

Different positions can be discerned within the creationist group: those that say that the natural world was created for human use, and, therefore, can be used in any way; another group believes the world was created for human use but thinks that human beings have no right to destroy it; and a minority which overtly expresses a conservationist view.

The first view is clearly stated by Anastasio Rodríguez, a Chol Indian: "When God created everything, he said 'Who is going to be in charge of all this?' So he said to Man: 'You are going to be in charge of everything, all I have created is yours, and nobody is going to take it away from you'. Man has everything in his hands". This last perception, that humans have everything in their hands, is actually held by few of the creationists: most believe that the conservation or destruction of Nature is up to God, not to men, and their attitude towards ecological issues is totally passive.

This view is also beginning to be turned into political discourses when arguing about the government ban against the cutting down of trees. One agricultural labourer stressed that "God made men to work, to take it easy and to command the world. God made Nature so we could command it. (Therefore) it is good to cut down the trees".

An extreme version of this is illustrated by the words of Rubén Merino, a shifting cultivator living in the Marqués de Comillas: "Well, what's the forest good for? It's just there. It should be good for something; otherwise it's just like having a calendar (picture). I want to work to leave something to my children, and if I have to strike down the forest, well, I'll strike it down". This does sound like the irresponsible view of a predator, but it is typical of those settlers who escaped from poverty elsewhere, are desperate to make a living, and are finding conditions deep in the rainforest harsh, unpromising and unmerciful, after having run a considerable risk and constant deprivation in everyday life.

This prevailing view that Nature belongs to humans, that the world was created for the benefit of men and women, is repeated over and over again, including among cattle ranchers and urban Palencanos, and, as yet, the issue of deforestation has not been taken up in this discourse. However, the environmental concern which does emerge quite frequently is anguish over the fact that "there are no more lands" in the sense that there are no more lands to expand into on which to grow food. This is the real issue for people in this region. A curious secondary issue which appeared in the interviews with urban Palencanos when asked what environmental problems they had was that of 'air pollution'—in an immaculately pure rural atmosphere! The explanation is simple: they watch television and are subject to the alarming concern over air pollution *in Mexico City*! An interesting subject for communications research: the issues that become public issues are those chosen to be communicated as such by the mass media. People will be more concerned about environmental issues in the city 850 km away and not able to *see* the devastation of the rainforest around them.

The second view is best illustrated in the words of the young wife of a policeman in Palenque who said "Jehovah made Nature so we could take care of it, to protect it and so that it will give us our nurture. We are a family with Nature, we are united" and added in a disapproving tone "there are people who are only interested in money and that is why they destroy everything, but in the beginning they will have their profit, but later they will have nothing". Note that this could be construed as a prototypical 'woman's' perception of the world. The view of the peasants within these groups can be illustrated by what Don Nicolás Díaz, a Chol Indian working as a mason in Palenque, said: "God made everything, plants, and Man, but Nature wasn't made for Man, it was made so it would reproduce itself, but now Man is doing away with it, he is destroying it very much".

Finally, the creationist but conservationist view brings together very different kinds of people, and the reasons they give in favour of conservation are very revealing. In general, fieldwork data showed that the perception of rainforest farmers in this group falls into two main categories: a majority who see the need to take care of the forest as a question of survival: "we live off the trees", a Tzeltal farmer said, "if it is depleted and the earth is left like a bare stone, we won't be able to get anything". Then there is a minority which makes a connection between the cutting down of trees and other undesirable environmental effects. A young Chol Indian said "if all the rainforest is cut down, the hills will plunge down; when the rains come it will all tumble down and nothing will grow any more".

More generally, interviews showed that the further away people's activities are from direct cultivation or plant or animal resources, the more conscious they are about the costs of environmental depletion. Obviously, they can afford to be more aware since they do not draw their livelihood directly from the forest, and they also have greater access to radio, television and magazines which carry such information. One example of this urban perspective was that of the owner of a local drugstore who said: "we need to have trees, they look so lovely, they give shade".

Differences are marked, in the use of terms, in tone and in information, between those holding a theistic view and those having a secular outlook. The latter are all white-collar workers, professionals, government officials and students. They do use the term 'Nature' and believe in the evolutionary process. "We are part of the evolutionary process. We are here to fulfil all the actions of humans, from being born to dying. But unfortunately, we don't live harmoniously with Nature", a government programme officer said.

The perception that humans are destroying nature is common among those interviewed, mostly because of "...a lack of awareness..."; "the obstacle (to protecting Nature)", added another government official, "is how beastly Man is. Many people cannot reason and if we don't reason, the future of society is destruction. The solution is to manage our participation with Nature to preserve it, to link awareness to reasoning". A man who paints signs in Palenque explained the problem by saying that "there is little awareness of the effects of the destruction of the rainforest. Everyone lives in a personal world, endangering our own species. Man has no consciousness of how this destruction is affecting him or of the fact that he is destroying. If Man has affected Nature, it is not because he dominates it, but because he has no consciousness".

Significantly, all such comments point at an individual lack of consciousness as the main reason the environment is being destroyed, that is, they attribute the destruction to a moral or intellectual handicap in human beings. None mentioned causes related to the economic or political system, or to demographic trends. It also becomes clear that they all blame others, while none of those interviewed considered that they themselves had any role to play in stopping this destructive process.

Finally, it is worth noting a single, idiosyncratic interpretation of the natural world in which a Chol peasant farmer, Amaro Martínez, declared that: "... the devil grabbed everything there is on earth and said it is his, and he doesn't like anyone to grab or kill animals because they are his... the devil feels he is the owner of the world but God gave us all the animals to eat, to have as food; if we kill many animals, God doesn't get angry but the devil does...".

This and other testimonies also point at something which is often forgotten in the mainstream, rather patronizing view of nature prevalent in urban culture. This is the customarily benign and bucolic view of nature by those of us living in such a domesticated environment that we don't realize that the rainforest, that is, what was called jungle just a few years ago, is also inhabited by poisonous snakes, predatory carnivores and a host of disease-carrying insects. Thus, it is important to remember that, for many of those who live it, and especially after nightfall, the rainforest also becomes a dark, devilish place.

14.7   Perceptions and Sustainable Behaviour

The content of beliefs and perceptions, as we have seen, can be analysed according to what we may call the 'intention of sustainability' towards the natural environment. It is quite another matter to try and explain how this cultural content is related to actual behaviour. As is well known, the relationship of thought to action is still one of the unresolved fundamental problems in the social sciences. Data from the research in Lacandonia can help us delineate a few working hypotheses on this.

In a recent paper, Rockwell (1992) has argued that culture in fact has little influence in environmental degradation, using Hinduism as a case in point. While this religion takes a notably respectful attitude towards animals and plants, this has not hindered the widespread depletion of natural resources in many parts of India. This may be true and the easy way out is to say that economic necessity at a given moment overcomes any religious or intellectual resistance; one could also argue that other values within Hinduism, mainly, those attributed to having children and large families, may override concern over Nature. In any case, it seems that no generalization, either arguing that culture has a predominant role or no role to play at all in altering behaviour towards the environment, can be demonstrated in a meaningful way.

The one problem which became evident in analysing the interviews in Lacandonia (and which was also present in previous research on culture and development carried out by the authors) is that the effects of cultural values on behaviour is practically impossible to assess in individual behaviour at a point in time. This is because individuals' behaviour patterns tend to alter according to what their neighbours are doing, and, most importantly, they are cumulative over time. To give an example, in the new settlements or *ejidos* of Marqués de Comillas, the landscape in the villages shows little difference from those outside Marqués in cases where Indian and *mestizo* (mixed) families are interspersed; however, in settlements colonized earlier, where Indians and *mestizos* live in different sections, the landscape tends to be very different in each section. Indians build their houses among the trees and surround them with fruit and flower plants; *mestizos*, who come from temperate zones, usually cut down all the trees, especially along the main streets. The point, then, is that cultural influences on environmental behaviour are mainly visible on a collective basis and in cumulative effects over time.

The comparative sustainability of each of these groups' *behaviour* towards the rainforest around them, though, can only minimally be ascertained in the verbal expressions of perceptions. It is mostly evident *in the combination of a number of actions towards their environment.*

Indian families—which are also internally differentiated as belonging mainly to the Tzeltal, Chol, Tzotzil, and Tojolabal cultures—tend to keep more trees in their immediate environment and in their maize fields, to use more rainforest and locally cultivated plants for food, medicine and decoration, and to have lower consumption needs and thus, fewer economic ambitions. Most Indians have also brought with them a centuries-old mythological and cosmological vision of the natural world surrounding them. Thus, one could generalize that they have a more 'sustainable' world view and behaviour.

Yet, on the other hand, Indian families have a higher population growth rate, and a more passive attitude towards family planning or ecologically-driven measures to save the rainforest. Also, they poach and sell wild animals in the same way as the *mestizos*. And, most importantly, at any given time, what makes their culture more 'sustainable' in intent depends on which of these components (agricultural methods, hunting and gathering practices, domestic production, population increase) in a given setting has the strongest negative impact on the natural environment.

This is to say that, seen in this light, the degree of 'sustainability' in their culture will not depend uniquely on cultural content but on how significant cultural traits combine in a system, and on the effects which these have in particular settings in time and space.

To summarize, the discussion on culture and sustainability will have to deal not only with the cognitive and cultural content of beliefs and perceptions but also with the effect these have on actual behaviour. We hope that the issues raised and the hypotheses suggested in this paper are useful in encouraging a debate which will be most important for the future of development.

References

Ayres, Robert, 1991: "Production of Goods and Services: Towards a Sustainable Future", Paper for the Symposium Humankind in Global Change: Indicators and Prospects, American Association for the Advancement of Science Congress, Washington DC, February.

Benería, Lourdes; Roldán, Martha, 1987: *The Crossroads of Class and Gender: Industrial Homework, Subcontracting and Household Dynamics in Mexico City* (Chicago: University of Chicago Press).

Bourdieu, Pierre; Passeron, Jean-Claude, 1970: *La Reproduction: Eléments pour une théorie du système d'enseignement* (Paris: Les Editions de Minuit).

Brundtland Commission, 1987: *Our Common Future* (New York: Oxford University Press).

CEPAL (Comisión Económica para América Latina), 1991: *El Desarrollo Sustentable: Transformación productiva, equidad y medio ambiente*, Vol. 146 (Santiago de Chile: CEPAL).

Fukuyama, Francis, 1989: "The End of History?" The National Interest, RAND Corporation, Summer.

Fukuyama, Francis, 1992: *The End of History and the Last Man* (New York: Free Press).

Mäler, Karl-Göran, 1990: *Sustainable Development*. Report of the Bergen Meeting on the Environment (Oslo: Norwegian Science Council).

Rockwell, Richard C., 1992: "Global Environmental Outcomes: The Current Scientific Understanding", Paper for the Population and Environment Workshop, DAWN-ISSC-SSRC, Cocoyoc, Morelos, 28 January–1 February.

Universidad Nacional Autonoma de Mexico (UNAM)

The National Autonomous University of Mexico (UNAM) was founded on 21 September 1551 under the name 'Royal and Pontifical University of Mexico'. It is the biggest and most important university in Mexico and in Ibero-America.

As part of the *Centennial Celebrations of Mexican Independence*, the National University was officially created on 22 September 1919. With the intention of widening educational opportunities in the country, the effort to launch the National University, though often hampered by adversity, was initially spearheaded by Congressman Justo Sierra in 1881. His vision finally materialized in 1910 with the inauguration of the *National Autonomous University of Mexico* at a ceremony held in the National Preparatory School Amphitheatre, where as Secretary of Public Instruction, Sierra told the audience that the thrust of the National University's educational project was to concentrate, systematize and disseminate knowledge in order to prepare the Mexican people for the future. One hundred years after the creation of the University, Justo Sierra's inaugural address still rings true: (…) *we are telling the university community today that truth is unfolding: go seek it* (…) *you have been charged with the realization of a political and social ideal which can be summed up thus*: *democracy and liberty.*

The fundamental aim of UNAM is to serve both the country and humanity, to train professionals, to organize and carry out research, mainly on national problems and conditions, and to offer cultural benefits in the broadest sense possible.

Website: www.unam.mx.

L. Arizpe, *Migration, Women and Social Development*,
SpringerBriefs on Pioneers in Science and Practice 11,
DOI: 10.1007/978-3-319-06572-4, © The Author(s) 2014

Centro Regional de Investigaciones Multidisciplinarias (CRIM)

The Regional Multidisciplinary Research Centre (CRIM) is an academic institution ascribed to the Coordination of Humanities at the National Autonomous University of Mexico (UNAM). It is located in the City of Cuernavaca on the Morelos Campus of UNAM. Its objectives are:

1. Focus on multidisciplinary research in social sciences, humanities and other disciplines, mostly aimed at tackling specific problems at the local, regional, national and international levels, and their implications within globalization processes.
2. Contribute to the creation of knowledge in relevant and innovative arenas addressing social problems that require the convergence of different disciplines for their study.
3. Contribute to the development of a multidisciplinary approach to humanities, and focus on the development of innovative theoretical and methodological perspectives.
4. Participate in educational programs so as to contribute to the academic training of professionals in social sciences, humanities and other disciplines.
5. Disseminate by all possible means the results of CRIM's research projects.

Website: http://www.crim.unam.mx.

L. Arizpe, *Migration, Women and Social Development*,
SpringerBriefs on Pioneers in Science and Practice 11,
DOI: 10.1007/978-3-319-06572-4, © The Author(s) 2014

About the Author

Lourdes Arizpe is a professor at the Regional Multidisciplinary Research Centre of the National Autonomous University of Mexico. She received an MA from the National School of History and Anthropology in Mexico in 1970 and a PhD in Anthropology from the London School of Economics and Political Science, UK, in 1975. She has pioneered anthropological studies on migration, gender, rural development, and global change and culture in Mexico, in Latin America, and in international research groups, both academic and policy-oriented. Professor Arizpe taught at Rutgers University through a Fulbright grant in 1979 and carried out research in India and Senegal with a John D. Guggenheim grant in 1981. She was director of the National Museum of Popular Cultures in Mexico 1985–1988. She was elected President of the National Association of Ethnologists of Mexico in 1986 and Secretary to the Mexican Science Academy in 1992. Professor Arizpe was Director of the Institute of Anthropological Studies at the National University of Mexico, was elected President of the International Union of Anthropological and Ethnological Sciences in 1988, and successfully organized its World Congress in Mexico in 1993.

Lourdes Arizpe became a member of the United Nations Commission on Culture and Development, and soon afterwards was designated Assistant Director General for Culture at UNESCO 1994–1998. She was elected President of the International Social Science Council for 2004–2008, and participated as a member of the Academic Faculty of the Global Economic Forum at Davos, Switzerland 2000–2004. At the United Nations Institute for Research on Social Development,

L. Arizpe, *Migration, Women and Social Development*,
SpringerBriefs on Pioneers in Science and Practice 11,
DOI: 10.1007/978-3-319-06572-4, © The Author(s) 2014

she was Chair of the Board 2005–2011 and a member of the Committee for Development Policy of the Economic and Social Council. She is a member of the Board of Trustees of the Library of Alexandria in Egypt.

Lourdes Arizpe became an Honorary Member of the Royal Anthropological Institute of the UK in 1995, and has received the Order of 'Palmes Académiques' from France in 2007, the Award for Academic Merit of the Universidad Veracruzana in Mexico, and an Honorary Doctorate from the University of Florida at Gainesville in 2010.

See also the websites on this and other books by Lourdes Arizpe: http://www.afes-press-books.de/html/SpringerBriefs_PSP_Arizpe.htm and http://www.afes-press-books.de/html/SpringerBriefs_ESDP06.htm.

About This Book

This book presents a selection of major research texts, published and unpublished, by Prof. Dr. Lourdes Arizpe Schlosser, a Mexican Pioneer in Anthropology. A global intellectual leader on culture, social development, sustainability, women's studies and indigenous groups, her texts provide both an outlook on the evolution of specific social scientific concepts and historical debates and a long-term and meta-analytical perspective integrating academic and policy discussions. By linking debates from different fields, the book helps readers to understand why people and groups make the choices they make and how the principles of social life must change to meet the challenges that new generations face in building social sustainability and effective environmental management in the twenty-first century.

L. Arizpe, *Migration, Women and Social Development*,
SpringerBriefs on Pioneers in Science and Practice 11,
DOI: 10.1007/978-3-319-06572-4, © The Author(s) 2014